Agricultural Mechanization in Developing Countries

STUDIEN ÜBER
ASIEN, AFRIKA UND LATEINAMERIKA

Herausgegeben vom Zentralen Rat
für Asien-, Afrika- und Lateinamerikawissenschaften
in der DDR
unter Leitung von Lothar Rathmann

Band 22

Recent Stage
and Prospects
of
Agricultural
Mechanization
in
Developing Countries

Recent Stage and Prospects of Agricultural Mechanization in Developing Countries

Proceedings of a Scientific Conference
of the Institute of Tropical Agriculture
and Veterinary Science
Karl-Marx-University of Leipzig
German Democratic Republic

July 1th – 10th, 1974

AKADEMIE-VERLAG · BERLIN
1976

Erschienen im Akademie-Verlag, 108 Berlin, Leipziger Straße 3—4

Lizenznummer: 202 · 100/105/76
Einbandgestaltung: Nina Striewski
Gesamtherstellung: VEB Druckerei „Thomas Müntzer",
DDR-582, Bad Langensalza
Bestellnummer: 752 894 5 (2152/22) · LSV 0305
Printed in GDR
DDR 34,– M

Foreword

The present volume contains the principal lectures presented during a scientific conference which was organized by the Institute of Tropical Agriculture and Veterinary Science of the Karl Marx University Leipzig as part of its 3rd International Summer Seminar (1. - 10. 7. 1974).

This conference was one of a series held after the founding of the institute (1960).

The problems of agricultural cooperatives in the developing countries were the topic of the first conference.[1]

In 1970, on the occasion of the tenth anniversary of the foundation of the institute a second conference was held, which dealt with ways of increasing production and the progressive transformation of agriculture in the developing countries, drawing on information from many branches of science.[2] The subject of the third conference was:

> "Recent stage and prospects of agricultural mechanization in developing countries."

There were good reasons for choosing this topic.

Contrary to view widely held by economists of the western world, agricultural practice has shown - particularly since the mid-sixties - that the

---

1 Cf., International Symposium on Agricultural Cooperation in the Developing Countries (lecture protocol), Leipzig 1965.

2 Cf., Problems of Progressive Agricultural Reconstruction in the Developing Countries (lecture protocol), Leipzig 1970.

mechanization of agricultural production in the developing countries is not a problem which is as yet remote, but one that demands immediate attention. A clear indication of this tendency is the fact that the rate of mechanization forecast by the FAO in the "World Plan for Agricultural Development"[3] has been exceeded by many developing countries and the material in hand for the 1975 FAO mechanization conference.[4]

Apart from this, it is becoming increasingly clear that food problems in the developing countries - which have become more acute recently - cannot be solved using traditional methods and means of production but must be solved by technological progress in the agricultural sector.

On the other hand, there is evidence that capitalist forms of mechanization in the developing countries are either not practical or are bound up with disastrous socio-economic consequences. In this connection, the positive experience gained by the socialist countries in the purposeful and planned mechanization of agricultural production on the basis of progressive social and economic changes in their agricultural systems, is of great interest to the developing countries.

There is no doubt that these aspects - as well as research done at the Leipzig Institute in the past years - helped to attract a large audience to this scientific conference. There were in all 147 participants from 24 countries included representatives of governments, farmers organizations, university lecturers, experts and young scientists.

Among the more prominent participants were the President of the All India Kisan Sabha, Dr. A. Z. Ahmed; the Syrian Vice Minister for Agriculture and Agrarian Reform Dr. N. Al Jumaa; the Chairman of the Egyptian General Organization for Milk and Meat Production, Prof. Dr. Ragab;

---

3 FAO, Provisional Indicative World Plan for Agricultural Development, Rome 1970.

4 Cf., Voss, C., Agricultural Mechanization, Production and Employment, FAO, Monthly Bulletin of Agricultural Economics and Statistics, Rome, Vol. 23, 1974.
FAO, Agricultural Production in the Developing Countries in Relation to the Targets for the Second United Nations Development Decade, Monthly Bulletin of Agricultural Economics and Statistics, Rome, Vol., 22, 1974.

the Director for Mechanization of the Egyptian Land Reclamation Organization, Dr. Hossary; the Deputy Director of the Department of Mechanization in the Agricultural Ministry of Ghana, Mr. O. K. Gyarteng; the Director of the Cooperative Department in the Agricultural Ministry of Somalia, Mr. M. Abokur as well as several deans and directors of institutes.

Participating in the work of the Seminar were delegations or individuals from Iraq, Syria, Egypt, Jordan, A.R. Yemen, India, D.R. Vietnam, Bangladesh, Burma, Sri Lanka, Tanzania, Somalia, Nigeria, Ghana, Guinea, Mali, Mozambique, Cuba and Chile. The socialist countries of Europe were represented by Bulgaria, Hungary, Poland, Czechoslovakia and the GDR.

The introductionary lecture (given by Prof. Dr. Michalski, Head of the Department of Agricultural Economics and Technology of the Leipzig Institute) outlined the opinion of the socialist agricultural economists with regard to agricultural mechanization in developing countries.

This opinion can be summed up in the following main points:

1. Agricultural mechanization is a necessary condition to overcome economic underdevelopment and food shortage and for the general economic growth in the developing countries.
2. Specific agricultural factors calling for mechanization are:
   - the need for reclamation of virgin land
   - the effective utilization of established projects (irrigation projects etc.)
   - the introduction of high yielding crops
   - the creation of new jobs for unemployed or underemployed people
   - the cultivation of areas infested by endemic animal diseases where draught animals cannot be used
   - the general need to raise labour productivity
3. Agricultural mechanization on capitalist lines is unsuitable for developing countries since this way of mechanization creates inevitably a monopoly of machinery ownership in the hands of the rural upper class and paves the way for exploitation, degradation and eviction of the

small farmers and tenants. In addition to that, a capitalist type of mechanization is a hindrance to planned national development of agriculture since private machinery ownership necessarily causes inequality, speculations and fluctuations in agricultural production.

4. From the economic and social point of view a planned, gradual mechanization concentrated on the main points of production and development and based on socio-economic reforms is the most suitable way of machinery utilization in the agriculture of the developing countries. Main points of production and development and based on socio-economic reforms is the most suitable way of machinery utilization in the agriculture of the developing countries. Main points of mechanization should be state farms, government projects cooperatives, centres of land reform as well as state or cooperative hiring stations.
5. A one-sided orientation on small-scale mechanization based on primitive technology (ox-ploughing, improved hand tools etc.) will not help to overcome economic underdevelopment. Under the conditions of the developing countries modern technology and the so-called "intermediate technology" should simultanously be introduced. Pilot centres of technological progress should be established to give practical orientation for gradual mechanization in the other parts of national agriculture.
6. Under the conditions of a non-capitalist reconstruction of the national economy a planned agricultural mechanization will not aggravate underemployment or unemployment but it will be able to create new jobs and in the long run it will help to redistribute human manpower step by step proportional to the various branches of economy.
7. The practical experiences of the socialist countries in Europe, Asia and Latin America proved the fact that agricultural mechanization based on socialist conditions is the historically shortest and most effective way to reconstruct agricultural production on progressive lines, to ensure full employment and to overcome underdevelopment and food shortage without the typical defects of capitalist agricultural development and mechanization.

8. Temporary difficulties of agricultural mechanization in the state farms or cooperatives of some developing countries are no reason to negate it but to analyse their causes and to overcome them by suitable technical or socio-economic measures.

For the three days discussion in all, 52 contributions had been received, of which 36 were actually presented. The remainder could not be heard due to lack of time, but some of these lectures have been included in this volume.

The discussion concentrated on various problems. A number of speakers dealt with political and economic aspects of mechanization (Dr. Ahmed, India; Dr. Jumaa, Syria; Prof. El Badrawy, Iraq; Mr. Gyarteng, Ghana; Mr. Abokur, Somalia; Mr. Bonora, Cuba; Mr. Bah, Guinea).

Managerial and technical aspects of mechanization in the developing countries were dealt with in particular by Dr. Hossary, Egypt; Prof. Vakarelski and Dr. Vassilev, Bulgaria; Dr. Pfeiffer, Dr. Glanze und Dr. Peipp, GDR; as well as Dr. Gunasena, Sri Lanka and Prof. El Fakhry, Iraq. Particular interest was taken in the lecture given by Dr. Glanze, who dealt with the results of research made at the El Gazair experimental station (Egypt).[5]

The lectures held by Dr. Jara, Dr. Oubrecht and Eng. Havrland, Czechoslovakia and Mr. Garcia, Cuba concentrated on the technical problems arising from the use of tractors and farm machinery in the tropics. The process of mechanization seen from the standpoint of natural science, agronomy, phytopathology and live-stock breeding, was discussed by Prof. Pagel, Prof. Franke, Prof. Fröhlich, GDR; as well as Prof. Ragab and Prof. Ali Kishen, Egypt.

Problems of agricultural training and research were dealt with by Dr. L.K. Ismail, Iraq; Dr. Neumann, GDR; Prof. Nowacki, Poland.

The general manager of "Agrotechnik" Leipzig, Dr. R. Hoffmann, described the experience of the agricultural machinery industry of the

---

5 A joint venture of the GDR farm machinery industry and the Egyptian General Organization for Utilization of Reclaimed Land

GDR with regard to the export and use of farm machinery in the developing countries.

The last day of the conference was devoted to lectures dealing with the present stage of development of mechanization in the agricultural production of the GDR. Six scientists and experts detailed the current transition to industrial crop and animal production in this country. This was followed by a visit to "agra 74", the agricultural show of the GDR at Leipzig-Markkleeberg.

Considering the wide range of participants and the different approach to the subject it is only too clear that different views were taken, which are not in keeping with each other in all aspects.

Therefore, the editor of this compilation does not share all opinions expressed in some lectures.

However, there was general agreement concerning the fact that a purposeful mechanization of the agricultural production conformed to the special circumstances in each country and based on socio-economic transformations is a burning necessity and acute task for the developing countries. For that the experiences made by the socialist countries are of great practical utility.

A three-days excursion programme took the participants of the Seminar to farm machinery plants at Neustadt (harvesting technology) and Weimar (soil cultivation machinery), to the cooperative association "Unstruttal", a mechanized milk production plant at Trossin, an agricultural chemicals centre at Querfurt and a nationally-owned stud farm at Graditz.

All those taking part in the Seminar stressed its great scientific and practical value and drew attention to the necessity of cooperation in this field, which is of considerable importance for the developing countries.

By publishing in this compilation the majority of the lectures and papers read during this conference the editor and the publishers comply with a wish vigorously expressed by the participants in the Seminar and many other interested parties. The present publication is to familiarize the public with some materials which will certainly enrich the international discussion on the very burning agrarian problems of the developing countries. This volume first of all applies to agricultural specialists, mechanization experts and economists in the developing countries. At the same time, we would like to address all those students, scientists, advisers and experts who deal in theory and practice with the agricultural development in the countries of Asia, Africa and Latin America.

The Institute of Tropical Agriculture and Veterinary Science thanks all lecturers and participants for their constructive cooperation and the Akademie Verlag Berlin for publishing this volume.

Contents

The authors

Prof. Dr. sc. K.-J. Michalski	Institute of Tropical Agriculture and Veterinary Science, Karl Marx University, Leipzig. Head of the Department of Agricultural Economics, GDR
Dr. Z.A. Ahmed	President of the All India Kisan Sabha, India (MP)
Dr. N. Al Jumaa	Vice-Minister of Agriculture and Agrarian Reform, AR Syria
Ing. E.P.G. Bonora	Lecturer, Faculty of Agriculture, Central University Las Villas, Santa Clara, Cuba
Dr. Ali Hossary	Director General, Department of Farm Machinery, General Organization for Utilization and Development of Reclaimed Land, AR Egypt
Ing. O.K. Gyarteng	Senior Agricultural Engineer, Ministry of Agriculture, Ghana
Prof. Dr. Ali A. El Khishen	Professor of Agronomy and Plant Breeding, University of Alexandria, Chairman of the High Committee of Mechanization, Ministry of Agriculture, AR Egypt
Prof. Dr. M.T. Ragab	Chairman, General Organization for Milk and Meat Production, AR Egypt
Dr. K. Ismail Laith	Head of the Division of Agricultural Mechanization, College of Agriculture and Forestry, University of Mosul, Iraq

Dr. M.M. Abdel Raouf	Institute of National Planning, Cairo, AR Egypt
Dr. S.A. Oni	Senior Lecturer, Department of Agricultural Economics and Extension, University of Ibadan, Nigeria
M. Abokur Haji	Director, Cooperative Department, Ministry of Agriculture, Somali Democratic Republic
Dr. A.K. El Fakhry	Director of the Agricultural Applied Research Centre, University of Mosul, Iraq
Dr. P. Glanze	Institute of Tropical Agriculture and Veterinary Science, Karl Marx University Leipzig, Department of Agricultural Economics, GDR
Dr. M.A. El-Deeb	Cairo University, Faculty of Agriculture, Department of Agricultural Economics, Giza, AR Egypt
Doz. Dr. L. Peipp	Institute of Tropical Agriculture and Veterinary Science, Karl Marx University Leipzig, Department of Agricultural Economics, GDR
Prof. Dr. sc. H. Pagel	Institute of Tropical Agriculture and Veterinary Science, Karl Marx University Leipzig, GDR
Prof. Dr. sc. G. Fröhlich	Director of the Institute of Tropical Agriculture and Veterinary Science, Karl Marx University Leipzig, GDR
Doz. Dr. Ing. Jára	Institute of Tropical and Subtropical Agriculture, Agricultural University Prague, CSSR
Doz. Dr. J. Oubrecht	Institute of Tropical and Subtropical Agriculture, Head of the Department of Agricultural Engineering Prague, CSSR
Dipl. oec. R. Hofmann	Director General, Farm Machinery Export Federation, Leipzig, GDR
Dr. R. Neumann	Institute of Tropical Agriculture and Veterinary Science, Karl Marx University Leipzig, Department of Agricultural Economics, GDR

Dipl. agr. G. Menzel — Institute of Tropical Agriculture
and Veterinary Science,
Karl Marx University Leipzig,
Department of Agricultural Economics,
GDR

Prof. Dr. E. Feder — Institute of Social Studies
Den Haag, Netherlands

Stage and Prospects of Agricultural Mechanization in the Countries of Asia, Africa and Latin America

by K.-J. Michalski

Whereas the technical details will be left to other speakers, I will turn my attention to those political and economic opinions, conditions and trends which influence the process of mechanization of agrarian production in the developing countries. I shall also attempt to put forward and discuss some suggestions for a well-planned gradual process of mechanization which is both efficient and socially progressive.

It is not our intention to dictate any formulas, but rather we want to make contribution to the international discussion of these problems. We want to suggest to our colleagues in the developing countries ways of overcoming the complicated problems of agrarian development, whereby we rely on the experiences of the successful mechanization of agrarian production in the socialist countries as well as on the results of our own research in many developing countries.

The problems of mechanization of agricultural production in the developing countries have long been the subject of sharp controversies, in the literature, at international conferences, in international organizations and above all in the developing countries themselves.

The exceedingly varied opinions which prevail in this field have very practical results in that they directly influence the formulation of official concepts of mechanization in the developing countries; they also influence planning and design activities, professional training and consequently the development of production in the agriculture of these countries.

The main contradiction within the framework of this international discussion exists between those progressive agrarian politicians, economists and mechanization specialists of the developing countries, the socialist countries and some far-sighted specialists from western countries who advocate a progressive, anti-capitalist mechanization policy, and the bourgeois agrarian specialists, who take a conservative and partly reactionary position toward the mechanization of agricultural production in the developing countries and do their best to promote private enterprize forms of machine utilization.

The opinions represented by bourgeois agrarian politicians and economists are quite varied as far as the technical details and methods of mechanization are concerned. They range from an open rejection of the use of agricultural machines in the developing countries and the recommendation of primitive forms of mechanization to a more or less undisguised promotion of the export interests of capitalist manufacturers of farm machinery.

The only point on which bourgeois experts agree is that agricultural mechanization in the developing countries should follow capitalist lines of private enterprise and that anticapitalist forms of machine employment should be prevented.

To begin with, permit me to characterize the different varieties of bourgeois concepts on the mechanization of agrarian production in the developing countries by giving some examples.

At the beginning, most bourgeois economists were of the opinion, that the conditions for the mechanization of agricultural production in the developing countries were not yet favorable and that only slow steps toward the improvement of the traditional tools were advisable. Only a few years ago, Prof. Denker, of West Germany, warned his students against the mechanization of agricultural production, describing it as "a sweet poison"[1] for the developing countries.

Gallwitz, in a paper on the employment of farm machinery in African countries, recommends the use of equipment similar to that "which was developed in small Southern German farms for drawing by cows after the first world war"[2].

In a standard text book, which is wide-spread in West German colleges, Schaefer-Kehnert also recommends "appropriate manual equipment" and writes, "the use of the scythe instead of the sickle would mean a step forward in some countries"[3]. Similar advice has been spread in the developing countries through many technical magazines.

K. W. Bean[4], the publisher of "World Crops", for example, opposes the use of modern farm technology especially by individual farmers in the developing countries and recommends an "improvement of existing methods" and the "modification of existing implements". Bean supports the misconception that it is not justified to introduce power machinery into the developing countries as long as the available human and animal power resources are not totally used. This means turning things upside-down.

It should not be difficult to understand that the complete utilization of human power resources in the agriculture of the developing countries has so far been impossible just because of the absence or shourtage of technology or other advanced means of production. This does not deter bourgeois economists from justifying theoretically such one-sided "recommendations" of primitive mechanization, which amount to the preservation of backward agrarian techniques in the developing countries.

Thus Priebe[5] bases his recommendations for retaining primitive forms of mechanization for a long period on a two-phase theory, which is modeled on the history of agrarian development in the industrial capitalist countries. In the first place, says Priebe, agrarian techniques in the developing Countries should "continue to be based on human and animal power and a modest improvement of implements", "without fundamental structural changes ... taking place in farms".

The actual mechanization of agrarian labour should only begin in a second phase, the beginning of which is not specified, with the transition to the use of driving machinery.

Prof. Smerdon, from the U.S.A., continued these thoughts and drew up a three-phase theory, according to which serious steps toward the introduction of modern farm technology into the developing countries could not begin until after 25 to 35 years[6].

It is obvious that these theories, which are directed against rapid economic progress for the developing countries, are to a high degree politically motivated.

Priebe emphasizes that any reversal of the mechanization phases which he favours could bring "great dangers" for the developing countries.

He sees "great dangers" in the "Russian successes" and continues, "the Russian collective appears as an ideal all the more for the oppressed small peasant in the Near East or the coloureds on the African plantations, the less they can free themselves from the fetters of feudalism and therefore hope to be liberated from the thousand year long hardship by communism"[7].

Bean expresses the fears of the international bourgeoisie as regards the social consequences of a progressive mechanization of agricultural production in the developing countries even more clearly when he writes in the above mentioned work: "Especially in peasant societies any change in the farming methods means a change in the way of life. It thus becomes a matter of social and often political revolution"[8].

Leaving aside the pro-colonialist and anti-communist terminology used here, these statements clearly show that the capitalist world and its bourgeois scholars view and utilize mechanization as a kind of "political weapon", with which to establish capitalist conditions of production in the agriculture of the developing countries.

Their goal is the gradual "evolution" of pre-capitalist forms of property and organization into a capitalist agrarian economy. At the same time they aim to prevent the establishment of modern, mechanized state projects and the effective combination of land reform and cooperatives with mechanization. These theories are designed to prevent public or cooperative pattern of agro-industrial integration, i.e. such forms of machine utilization which are the economic basis of social progress.

It is not to be overlooked that bourgeois theories of mechanization have had and continue to have considerable influence on the practical mechanization policies of the developing countries.

For instance, the agrarian development of India and other Asian countries for almost 20 years dominated by the concept of the so-called "bullock economy". Innumerable Indian publications proclaimed the 70 million bullocks in the country as the main or only power source of agriculture and, at the same time, warned against the "dangers of mechanization", which would allegedly bring uncontrollable erosion, the slaughter of cattle, increased unemployment etc. As a result, much money and thought was used to develop, produce and use the simplest animal-drawn equipment.

The experience in the past years has been that this concept did not solve the country's agrarian and food problems. Since 1967, the mechanization of Indian agricultural production has developed more rapidly than expected but on a socio-economic basis which excludes most small farmers and tenant farmers from the benefits of technology.

The official plans and agrarian programmes of many African countries also contain mechanization concepts which follow these bourgeois recommendations and theories.

The transition from the hoe to the ox-plow is frequently presented as the main approach toward an increase in production and modernization of farming methods. As a result, considerable sums and human resources are spent on the development, production and application of the simplest animal-drawn equipment in these countries.

Although this aspect is of some importance in the countries of tropical Africa, it must not be overlooked that an effective modernization of production methods and a lasting increase in yields and productivity are not possible in this mannor. Even if the majority of peasants in an African country were to change over from the hoe to ox-drawn equipment, the country in question would still have a backward agriculture, which could not contribute to the consistent development of the economy as a whole.

Bourgeois mechanization theories likewise affected the planning and consulting activities of international organizations.

Thus the FAO "Indicative World Plan for Agricultural Development", which was published in 1970, assumed a "relatively slow progress"[9] of mechanization of agricultural production in the developing countries and thereby underestimated the rate of tractor and machine utilization in these countries. The FAO foresaw for the period from 1965 to 1975 an increase of the tractor population[10] in the developing countries from 544.000 to 1.005.000[11].

By 1971, the developing countries had in fact already gone beyond these forecasts and had reached at that date a number of 1.046.800 tractors.[12]

In some countries, for example in north-west Africa, even the IWP forecasts for 1985 were left far behind.[13]

In view of the present annual growth-rates, the construction of domestic tractor works and as a result of the revision of mechanization programmes in many developing countries and the increasing investment power resulting from higher prices for raw materials, it can be assumed that these countries will have reached or exceeded a population of ca. 1,5 million tractors by 1975.

This produces a greatly changed state of affairs for the 1975 - 1985 period.

Table 1 shows the growth rates of the tractor numbers of the developing countries from 1965 to 1970 as compared with IWP objectives.

Table 1 Growth Rates of the Tractor Numbers of the Developing Contries as compared with the IWP objectives of the FAO (% annually)

Region	actual rate of growth	FAO objectives	
	1965-1970	1965-1975	1975-1985
Africa (south of Sahara)	6.2	4.8	4.8
Far East			
2-wheel tractors	28.0	19.1	20.0
4-wheel tractors	15.6	12.5	13.2
Near East	18.8	7.3	5.1
North-West Africa	6.0	3.1	2.9
Latin America	3.8	4.3	3.6
Total	9.5	8.4	9.3

Source: FAO, Monthly Bulletin of Agricultural Economics and Statistics, Rome, Vol. 22, 1974, 4, p. 13.

Behind these averages, we find widely differing developments in the countries of Asia, Africa and Latin America. While in many countries of tropical Africa and in some Asian countries the extent of tractor and farm machine employment is still very low, in numerous other developing countries an irreversible transition has occurred from traditional subsistence farming to modern, intensive methods of agrarian production, which will lead to a progressive increase in the use of modern means of production. There will be a further increase in the utilization of tractors and farm machinery to match the introduction of high-yielding crops, agricultural chemicals, irrigation, processing industries etc.

A case in point is India where the use of agricultural machines and tractors increased very slowly up to the middle of the sixties. Up to that time the available machines and implements remained extremely backward and far short of the demand. According to the 1961 census, only 58 % of all Indian peasant families owned a wooden plow and only 3, 4 % owned iron plows. Only 0, 6 % of all farms owned a pump, while every sixteenth village had a tractor.[14] Since 1966/67 the situation has drastically changed. The so-called "Green Revolution" produced a complex process of intensification, starting with the cultivation of high-yielding varieties (wheat, rice), and immediately resulting in the increased use of modern means of production, especially fertilizer, irrigation works, plant protection, as well as tractors and machinery. Since then, the number of tractors in India increased by leaps and bounds - at the beginning mostly through imports, now mainly through India's own production.

Table 2 illustrates the rapid increase in Indian tractor numbers in the last few years.

Table 2 Development of the Number of Tractors in India (1946-1973)

year	number of tractors
1946	4, 524
1951	8, 635
1956	21, 005
1961	31, 016
1966	54, 021
1971	143, 000
1972	173, 000
1973	ca. 200, 000[x]

[x]estimate

Source: Jain, B.K.S., Prospects of Farm Equipment Industry, Agric. Mech. in Asia, Tokyo, Vol. IV, 1973, 2, p. 95.

Of great economic importance is the fact, that in recent years India has built up its own tractor and farm machinery industry, which provides most of the equipment used in the country. By 1973, seven tractor plants with an annual licensed capacity of 60, 000 tractors had started production. At present about 20, 000 of different sizes between 20 and 50 hp are produced annually. The production of 2-wheel tractors has now begun, for which a total capacity of 40, 000 per year has been licensed. Moreover, the Indian farm machinery industry produces at present about 30, 000 threshing-machines annually as well as a considerable number of Diesel engines, electric motors, plant protection units, disk harrows etc. The production of small combine harvesters has begun.[15]

Egypt is a developing country, where the use of tractors and agricultural machines also has made great progress. While the employment of tractors on a cooperative basis and integrated crop rotation is furthered in the traditional areas of peasant settlement in the Nile delta and Nile valley, state agricultural enterprises arose in the newly reclaimed land areas on about 300, 000 ha which have achieved a high degree of mechanization.

According to informations by Egyptian authorities, the mechanization of important agricultural operations is developing as follows:

Table 3 Degree of mechanization of agricultural production in Egypt (%)

	single farms		state farms on reclaimed land	
	1972	1978ˣ	1972	1978ˣ
Ploughing	60.0	100	100	100
Seed-drilling	0.1	30	65	100
Plant protection	80.0	100	80	100
Grain harvest	72.0	100	100	100
thereof				
Combine harvest	0.0	30	55	100

ˣPlan

Besides these strides in the mechanization of plant production, Egypt has undertaken far-reaching steps towards setting up highly mechanized farms for animal production, towards extending its agricultural processing industry and the development of its own tractor and farm machine industry.

These trends in the mechanization of agrarian production - illustrated here by the example of India and Egypt - have also emerged more or less distinctly in many other developing countries.

This is indicated also by the continuing construction of home industries for tractors and agricultural machinery in the developing countries.[16]

Complete tractor industries which largely satisfy home needs, have also been established in Mexico, Brazil and Argentina, whereas Egypt satisfies a part of its tractor requirements through home production. Assembly plants for tractors have been erected in Syria, Pakistan, Thailand, Sri Lanka, Iran, Turkey, Algeria, Marocco and other countries. In these and many other developing countries - among them African countries like Tanzania, Senegal, Cameroon, etc. - the production of tractors - or animal-drawn equipment has been started, and some developing countries are already exporting agricultural implements.

This survey of the mechanization trends of the last 6 to 8 years in the developing countries clearly shows that the above-mentioned theories and recommendations of bourgeois economists and politicians have been increasingly refuted in pratice. More and more developing countries are reaching a stage in which they overcome the traditional backward production methods and achieve scientific and technical progress on the sphere of agriculture.

We cannot subscribe to the view expressed in many places and also by representatives of the FAO that this is an "undesirable" development.[17] We consider this a desirable and urgently needed development of agrarian productive forces in these countries, which will contribute to an increase of food production and a consolidation of national economic mechanisms of reproduction which were deformed by colonialism and neocolonialism.

The further development of the mechanization process will surely prove that the bourgeois forcecasts were false not only in regard to the rate of mechanization, but also that the capitalist forms of machine employment, which presently prevail in most developing countries, are not suitable.

Privately financed, capitalist mechanization is a spontaneous process, the ownership of machines - and the political and economic hegemony that goes with it - rests in the hands of a small group of large farmers, landowners and capitalist firms and accelerates at the same time the process of capitalist polarization, which leads to the ruin and expulsion of small farmers and tenants on a large scale.

Permit me to illustrate some ideas on the mechanization of agricultural production in the developing countries from the standpoint of socialist agricultural economy.

Socialist economists have always criticized the capitalist approach to the mechanization of agricultural production and revealed its contradictions and negative socio-economic consequences.

Karl Marx and Friedrich Engels thoroughly analyzed the economic and social aspects of the use of machinery - and of farm machinery - in capitalism, and Lenin discussed the problems of machine employment in capitalist agriculture at great length in numerous papers.[18]

Marx proved that the historically necessary mechanization and industrialization of agriculture under the conditions of a capitalist mode of production becomes a "martyrdom of the producer" because the means of production are at the same time "means of enslaving and exploiting ... the labourer"[19] and because this involves the "expropriation of large masses of the agricultural population" and "changing the dwarf-like property of many into the vast property of a few".

Marx writes, that the expropriation of the agricultural producer is carried out with the most ruthless vandalism and under the pressure of the most filthy and petty passions.[20]

Lenin arrives at a similar conclusion and writes that capitalism raises farm technology to a higher level and advances it but this it can do only by ruining, strangling and crushing the mass of the small producers.

These forecasts made by the classical authors of scientific socialism have been verified a thousand times in the agricultural history of the last decades.

The rapid progress of farm technology in the highly developed capitalist countries meant million-fold ruin for small and medium-sized farmers for the benefit of the large capitalist agrarian enterprises and to the advantage of industrial and banking capital. Thus the number of farms in the U.S.A. decreased from 6.5 million (1930) to 2.8 million (1972) and it is expected that in 1980 only 1.5 million farms will survive.

From 1960 to 1970, not less than 2.25 million farming families were forced to give up their farms in six capitalist countries of Western Europe (Great Britain, France, West Germany, Italy, Denmark and Norway)[21], primarily because they could no longer bear the financial burden of private mechanization. In the last few years, similar processes have started in the developing countries where the mechanization of agriculture follows capitalist principles. In connection with the so-called "Green Revolution" and the enforcement of bourgeois land reform laws above all in Asian countries, millions of small farmers and tenants lost their plots which are being merged into large capitalist holdings. In addition to many other reports this is also confirmed by a United Nations study[22] which is based on investigations in Pakistan, India, the Phillippines and Sri Lanka and in which the fact is established that more and more tenant farmers are becoming proletarians, whereas the absentee landlords are becoming capitalist farmers, i.e. agrarian businessmen. In this way, millions of people are being deprived of their means of existence and driven into the big city slums, since most developing countries will not be in a position to give them alternative work in the non-agrarian sector in the near future.

Together with leading figures and scientists from the developing countries we have repeatedly warned against the far-reaching social and economic consequences of such spontaneous and haphazard mechanization and intensification of agriculture in the developing countries, which is directed solely toward the profit and power interests of a small capitalist stratum.[23]

It is not the tractor, not the machine, not the high-yielding varieties and mineral fertilizer which ruin the small farmer and tenant, which deprive him of his means of existence, grade him to a slum-dweller, but rather the private, capitalist form of these intensifying processes, which convert the "dwarflike property of many into the giant property of a few".

If the inevitable processes of mechanization, intensification and reducing the agricultural labour force do not take a spontaneous private capitalist form, but rather the form of cooperatives assisted by systematic planning provided by a progressive state, i.e. in an anti-capitalist manner, negative socio-economic consequences can be avoided and it will then be possible to convert the "dwarf-like, backward small property of many into the giant collective property of all".

The agrarian history of the socialist countries has clearly proven the realistic quality and the advantages of this approach.

The basic guide-lines for the agrarian mechanization policies of the socialist countries go back to Karl Marx, Friedrich Engels and W.I. Lenin. Marx spoke of the necessity of replacing the backward, small farmstead with the "scientific ones ... through the union of agriculture and industry"[24], in order to produce a technological and economic balance of industry and agriculture. Marx writes, that at a certain peak of industry this disporportion must decrease, i.e. the productivity of agriculture will increase more quickly than that of industry.

In April 1917 - foreseeing the coming socialist revolution - Lenin favored transition to an efficient system of production on a large scale, since the system of small holdings is not capable of freeing humanity from mass poverty and oppression.

In December 1918, 11 months after the victory of the October Revolution, at a time when the young Soviet Republic still had great economic problems, Lenin drew up a detailed programme of socialist development of agriculture, commenting that the miracles of technology must primarily be used to transform the industry which is of the greatest concern to the majority, which employs the majority and which is most backward, i.e. agricultural production.

At that time Lenin dreamed of "100,000 first-class tractors"[25] and in 1921 initiated a decree about the construction of a soviet tractor and farm machinery industry.[26]

The socialist countries have followed these guide-lines and built a modern, highly mechanized socialist system of agricultural production.

The first example and the basic model of a mechanized socialist agriculture was created in the U.S.S.R.

By establishing state farms, by merging 24 million backward small holdings into large collective farms, by building a viable tractor and farm machinery industry and by systematic state management of agriculture, the U.S.S.R. created the powerful technological base and the progressive conditions of production which made it possible to mechanize completely or almost completely all basic farming operations, change over to industrial methods of production and reclaim vast areas of land. By reclaiming 55 million ha of virgin land from 1954 to 1965, the U.S.S.R. brought an area under cultivation which is almost as large as the combined areas under crop in Turkey, Iraq, Syria, Egypt, Algeria and the Sudan.[27]

The rapidly progressing intensification of agricultural production, especially in the last few years, has meanwhile enabled the U.S.S.R. (1973) nearly caught up to the U.S.A. in the field of wheat and cotton production and to become the world's largest sugar producer.[28]

This rise in production will surely continue at a quick pace in the coming years, since the U.S.S.R. is presently making large agricultural investments.[29]

This progress is based to no small measure on the technical strength of Soviet agriculture.

There were in the Soviet Union in 1973, 2.1 million tractors, 656,000 combine harvesters and a wide range of specialized machinery, means of transportation and technical installations.[30]

This machinery enables a high degree of mechanization, and the plowing, sowing and harvesting of wheat and silage crops have been completely mechanized for many years. Complete mechanization is now envisaged of all other operations in plant production and extensive mechanization of animal production is planned. It is important to remember that the mechanization of Soviet agriculture has been complicated to a high degree by various historical circumstances such as the economic backwardness of old Russia, the necessary priority of industrial construction, the enormous human and material losses caused by the civil war and the Second World War, the heavy economic burdens the Soviet Union had to carry as a result of the "cold war" and the aid given to other socialist countries and many developing countries.

Using the Soviet experience and assistance and benefiting from favorable historical conditions, the other socialist countries introduced mechanization in a shorter period and with fewer difficulties. Nonetheless, it was a great effort for the socialist countries created after the Second World War to mechanize their agrarian production.

A case in point is Bulgaria, which was in an economic situation similar to that of many developing countries up to 1944-1945, but which established a highly mechanized, socialist agriculture over the relatively short period of 20 years.

Up to 1944, Bulgarian villages were characterized by poverty, backwardness and exploitation of the farmers.

Of 1.1 million farms, two thirds were smaller than five hectares. The arable land was divided into 12 million small holdings. About one fourth of all farmers were tenants or share croppers. The great majority of farmers worked with wooden plows and other primitive instruments. 25 % of all farming families had no draft animals and the country had only 1,500 tractors which belonged exclusively to the big farmers.[31]

This situation was transformed by the land reform of 1944-45, by the establishment of cooperative farms and state farms and, since 1970, of agro-industrial complexes[32]. Today there are 860 cooperative farms and 150 state farms, which together form 161 agro-industrial complexes.[33] This has made all-round mechanization of agricultural production and the transition to industrial methods possible. By the middle of the sixties, the full mechanization of all principal operations in plant production was achieved.

Table 4 Degree of Mechanization of Agricultural Operations in the People's Republic of Bulgaria (%)

	1952	1956	1965	1967	1972
Ploughing, total	64.5	84.6	96.4	97.8	
Deep autumn ploughing	86.2	99.9	99.4	99.5	99.1
Harrowing	43.4	81.9	98.4	98.0	
Sowing	34.5	75.5	95.3	97.1	99.5
Harvesting	42.3	44.4	92.5	96.6	99.2

Source: Roussinow, S., Economic Development of Bulgaria after the Second World War, Sofia 1970, p. 98.

This high degree of mechanization, which is also being introduced in animal production and in fruit and vegetable cultivation, was established without the expulsion of small holders, without spreading unemployment and other symptoms of capitalist intensification. It has brought unprecedented prosperity to the rural population and benefited the Bulgarian economy as a whole.

Mechanization took a similar course in other socialist countries. Table 5 shows the number of machines in G.D.R. agriculture.

Table 5 Degree of Mechanization in GDR-Agriculture

	1960	1970	1973
No. of tractors	70,566	148,865	143,266
arable land per tractor (%)	90	40	41
No. of combine harvesters	6,409	17,911	11,957
grain area per combine harvester (ha)	360	123	193
No. of potatoe combines	6,386	12,000	10,193
potatoe area per combine (ha)	114	52	60
No. of sugarbeet combines	3,665	5,276	4,896
sugarbeet area per combine (ha)	65	36	47

Source: Statistische Jahrbücher der DDR, 1965 und 1974

This systematic increase in the number of machines enabled the mechanization of all basic operations in plant production. Table 6 illustrates this for harvesting operations.

Table 6 Degree of Mechanization of Harvesting Operations in GDR-Agriculture (% Area under Crop)

	1960	1970	1973
grain harvest by combine harvesters	37.9	98.9	99.2[x]
potatoe harvest by combines	24.6	76.7	89.3
sugarbeet harvest by combines	46.1	91.4	99.1

[x]1972

Source: Statistische Jahrbücher der DDR, 1965 und 1974

An important aspect of machine use in the agriculture of the G.D.R. is the systematic replacement of existing machines by newer and more efficient models. The establishment of "Cooperative departments of Plant Production" between neighbouring cooperatives and state farms encourages this process. For instance, the large Soviet tractor K 700 (215 hp) is being used for heavy field work in place of the usual middle weight tractors (60-90 hp).

In the past six years the modern, highly efficient combine harvester E 512, which harvests an average of 350 hectares per campaign, replaced the combine harvester E 175, which accomplished an average of ca. 150 hectares of harvest area per campaign. In 1973, 89.3 % of the wheat in the G.D.R. was harvested with modern combine[34].

Mechanization has made rapid progress in animal husbandry and has alleviated such strenuous operations as milking, litter handling, fodder handling and storage processes. Recent planning provides for the gradual introduction of large-scale industry-type animal production establishments. These facilities use the most advanced equipment and have the following average capacities[35]:

Dairy farming	1,930 animals
Calf raising	1,600 - 3,200 animals
Young cattle raising	4,480 animals
Cattle fattening	16,000 animals
Pig raising	5,600 animals
Pig fattening	25,000-100,000 animals
Egg production	200,000-400,000 animals
Broiler production	up to 3 million broilers/year

Such industry-type plants now accomodate some 6.5 % of the farm animal population of the G.D.R.

Developing countries will most probably be able to profit from the experience of Cuba where great efforts have been made to mechanize farming operations.

In Cuba, state farms cultivate some two thirds of the crop area.[37] The remaining one third is divided among private farmers who are now beginning to work together in cooperatives.[38]

These new socialist relations in production emerged after the 1959 revolution and have substantially advanced farm mechanization. As early as 1960, the Cuban government adopted a long-term programme for the import of tractors and agricultural machinery providing for the purchase, by the year 1972, of 68,704 tractors, primarily from the U.S.S.R. and other socialist countries.[39]

By the year 1970, Cuba had also imported the following[40]:

1,270	sugar cane harvesters
9,150	tractor loaders
15,000	sugar cane carts
1,450	harvesters (especially for rice)
56,000	plant protection units
3,000	fodder harvesting machines
380	milking machines and milking plants

Efforts are currently being made to step up mechanization, and Cuba is beginning to make its own farm machinery such as plows, disc harrows, drilling machines, fertilizer distributors and sugar cane harvesters[41].

There were three principal areas of agricultural mechanization in Cuba - sugar cane production, rice farming and the production of milk and meat. Sugar cane cultivation and husbandry operations are now almost completely mechanized. The hard and time-consuming job of loading harvested sugar cane into carts was first dealt with, and 90 % of the work is now done by tractor loaders. The cutting of sugar cane remains the basic problem, and this is still mostly done by hand. The answer is the cane harvester which is being developed by Cuban and Soviet experts. In view of the progress made so far[42], it may be expected that harvesters will become a more common sight on Cuban fields in the near future.

Great strides have been made in mechanizing rice production which has been extended to an area of some 200,000 hectares[43].

In 1971, the mechanization, in per cent, of rice production operations was as follows[44]:

Tilling	100 %
Sowing	50 %
Plant husbandry	50 %
Harvesting (with harvester-combines)	90 %

With new equipment being added according to plan, full mechanization of rice production may be expected in a few years time.

More examples could be given of successful mechanization efforts in European socialist countries, and in the Korean and Mongolian People's Republics. The large-scale use of machinery is only just beginning in the Democratic Republic of Vietnam, after the end of the brutal American bombing raids.

The example of the socialist countries shows that agricultural production can be re-structured, and a high degree of machinery employment and productivity achieved, in a historically short period. This is possible even for countries with widely differing starting positions, without giving rise to social diversification into rich and poor people, without unemployment, without ruining small farmers and tenants, and without disastrous fluctuations in agricultural output.

The experience of socialist countries, and especially those countries which were economically backward only a few decades ago, gives the lie to bourgeois arguments that conditions are not ripe in the developing countries for mechanizing agricultural production.

A number of factors can be listed which necessitate the use of tractors, farm machinery and processing facilities in the developing countries, as follows[45]:

1. Most developing countries have a vast potential of virgin land which can only be cultivated if tractors and machinery are used. This land cannot be cultivated with the conventional implements in use in these countries.
2. Most developing countries expend large sums for irrigation projects and making new land available, and these projects will not be economical if old farming techniques are retained. On the other hand, optimum results can be expected with the use of advanced techniques and equipment.
3. High-yielding varieties which have been used in developing countries on a large scale recently, require advanced tilling and plant husbandry methods, modern harvesting techniques and higher expenditures for transport, storage and processing. These can be realized only with the help of modern technology.

4. Draft animals cannot be used in many areas in the developing countries because of the danger of sleeping sickness and other endemic diseases, or their use is severely restricted. This calls for the employment of tractors and farm machinery.
5. Measures aimed at intensifying farm production are necessary in those areas in the developing countries where there is a high population density, a shortage of land and excess labour. Such policies will increase productivity per acre and create more jobs. Possible approaches would be irrigation, the erection of processing facilities, the use of herbicides and other chemicals, better tilling methods and the development of animal husbandry etc.[46].

In this connection, permit me to comment on the widespread argument that agricultural mechanization in the developing countries would aggravate the problems of underemployment and unemployment. There is no question that population density and the labour situation are important aspects which need to be considered in the planning and implementation of mechanization programmes in the developing countries. However, it would mean a complete reversal of logic to argue that mechanization neccesarily leads to underemployment and unemployment.

It is well known that unemployment and underemployment have been so widespread in the developing countries not because of mechanization but because of a lack of it and a shortage of machinery and other means of production which make human labour more efficient.

For example, many peasant families in tropical Africa are underemployed not because they use machines but because the hoe is their main implement. This obsolete production method limits their capacity to 1 - 2 hectares of crop area and keeps yields at a low level. Many peasant families in Asia and the Arab countries are underemployed not because they use machinery but because their plots are extremely small and the wooden plow limits productivity to a minimum. This is why a high productivity of human labour and full employment cannot be achieved by preserving these backward relations of ownership and obsolete farming methods but by abandoning them.

This applies not only to agriculture. Relative overpopulation and the related underemployment in many areas of the developing countries affect the economy as a whole. They are symptoms of a general state of economic underdevelopment and can only be overcome if the national economy as a whole is developed[47]. This calls for industrialization and a reform of agriculture, for the development of crafts, trade and services, and for improving education and the health system. Economic growth and the balanced development of the entire national economy are the only means of releasing excess labour from the agricultural sector to other industries. This will finally put an end to underemployment and joblessness.

These will be basic considerations in defining the role that mechanization is to play in a developing country, or in one of a number of regions with differing economic and other features. It should be noted, in this connection, that farm machinery can be used to relieve manual labour - this is the way it is now used in the socialist countries - but also, at least for a limited period, to create new jobs. Which of these two approaches is selected will depend on the economic situation in a given country or region.

In some countries or regions where large areas of unused land are being utilized and there is a shortage, rather than an excess of labour, mechanization will obviously have to save manual labour.

In densely populated areas with excess labour machinery should be used to create a maximum of new jobs. This can be done by applying mechanization to the following fields:

- Irrigation and draining
- Soil conservation and land improvement
- Improved tilling
- Plant protection
- Processing
- Electrification
- Transport and handling

It has been found in many densely populated developing countries such as Thailand, Taiwan, the Philippines, and in parts of India, Egypt and

Kenya that a concentration of machinery in the above mentioned fields resulted in a much better use being made of human labour for a given area of land, and in a substantial increase in production. [48]

In view of our experience and summing up what has been said so far, we believe that the developing countries should adopt a policy of planned and gradual mechanization in selected fields of agricultural production, based on socio-economic reforms.

In order to avoid the disadvantages of mechanization under capitalist conditions, or to keep them to a minimum, technological and scientific progress must be matched by socio-economic advance. The employment of machinery must be an integral part of a broad plan designed to increase production.

Large-scale government projects, state farms, land reform centers, cooperative farms and state-run or cooperative machine hire stations would seem to provide a favorable basis for mechanization in the developing countries.

The fact that we suggest planned and gradual mechanization in selected fields does not mean that we propose attempts to achieve full mechanization in a short time. Such attempts would ignore the economic and labour situation in the developing countries und require a higher level of education of the rural population.

The introduction of improved animal-drawn equipment and other forms of small mechanization will probably be a necessary transitional stage. However, while this may apply to some developing countries, this can never be the sole or principal way of using agricultural technology, or increasing production. Our concepts start from the assumption that both the advanced technology and what is sometimes described as "intermediate technology" should be employed in parallel. Local conditions will then be decisive in planning the scope of employment of either equipment.

Khalatbari refers to this as a "combined approach" and speaks of a kind of "coexistence of farms using different forms of technology"[49].

However, this "combined approach" can be successful implemented only under the conditions of a non-capitalist way of development. It means that advanced equipment will be introduced in selected enterprizes (state farms, state farming projects, cooperatives), or in some geographical areas or branches of production where there are favorable conditions for mechanization. More simple forms of mechanization will be used in other areas until advanced technology can take over.

This concept is in some ways reminiscent of "selective mechanization"[50] as propagated by the FAO but in some ways it differs from the FAO approach. The latter is mostly defined in terms of mechanization in a private enterprise economy covering selected operations (such as plowing, plant protection, small-scale irrigation etc.). The equipment proposed for this is suitable for medium-size or large farms. The FAO approach would seem to extend indefinitely the period in which intermediate technology is used, with no specific time set for the introduction of more sophisticated equipment. This policy would also mean that mechanization serves the development of a production system based on private enterprise and capitalism.

In contrast to this, we suggest the establishment, as soon as possible, of centers of mechanized large-scale farm production. This will enable individual countries to profit from international developments in science and technology. Such establishments would be testing and breeding grounds for experts, and they would serve to demonstrate the advantages of production on a large scale to the farmers of a country. They would represent a "breakthrough" toward the employment of modern technology and could be used for seed growing and stock breeding and provide advisory and training services. One example of a mechanization policy along these lines is Tanzania's second five-year plan which speaks of a "balanced programme of mechanization". Under this programme, tractors and heavy machinery will be concentrated in carefully selected areas in such establishments as state farms and cooperatives. Mechanization for the individual farmer is provided in the form of animal-drawn equipment and by encouraging machinery and tractor hiring cooperatives[51].

Such policies combine a suitable form of ownership of technology with adequate facilities for planning machine employment. State farms, cooperatives and machine hire stations run by the state or by cooperatives have been found to be well suited for a planned approach to selective mechanization in the agriculture of developing countries. Capitalist forms of machine ownership and use are thus effectively restricted, and farmers avoid the excessive cost of individual mechanization. The fact that the individual farmer has to carry the heavy burden of equipment costs is a major reason for the continuing ruin of small and medium-size farms in the capitalist countries.

In West Germany, the equipment used by a 15-hectare farm represents an average capital value of some 50,000 Marks.[52] There is economic pressure on these farms to spend more than is necessary on tractors and agricultural machinery, and this is often presented as symbolizing economic prosperity. The truth is that farmers are highly indebted and many have been forced to sell their farms. The average indebtedness of farms under 20 hectares rose from 740 DM/hectare in 1958 to 2,847 DM/hectare in 1971. For farms between 20 and 50 hectares, this figure increased from 792 DM to 2,692 DM over the same period.[53]

This means that a 15-hectare farm carries an average debt burden of some 43,000 DM and has to pay about 4,000 DM interest per year. The result has been that since the mid-1960s reinvestment in the under-50-hectare group has been exclusively financed by credits and that the debt burden grows faster than capital investment in these farms.[54]

The total debt burden on West German agriculture increased from 3.7 thousand million Marks in 1949/50 to 32.7 thousand million Marks in 1972/73 while interest charges rose from 235 million Marks to 2.2 thousand million Marks over the same period.[55] It will be appreciated that a comparable burden on farms in the developing countries would be highly undesirable and uneconomical.

Similar disadvantages would result if private firms were allowed to hire out farm machinery and tractors to farmers. Such undertakings would be managed according to capitalist principles and would seek maximum profits through the highest possible rental charges. This would make their services prohibitive to the majority of farmers in the developing countries, or it would mean severe financial burdens. Apart from this, farmers would become dependent on these firms in a way which is not much different from their dependence on private money-lenders or traders.

Considerations such as these have led some developing countries to concentrate machinery which is to be hired out in state farms, cooperatives and machine hire stations run by the state or by cooperatives, rather than leaving this field to private enterprise. Both profitable results and problems have been reported as a result of such policies.

Our own investigations in a number of developing countries indicate that good results have been obtained especially in these areas[56]:

1. Extension of crop area
2. Higher yields
3. Increased production of export crops and higher foreign exchange earnings
4. Higher incomes for farmers and farmworkers
5. Farmers will be more prepared to use modern production methods (fertilization, plant protection, crop rotation, transition from migratory to permanent crop farming)
6. Emergence of new forms of production (group farms, cooperative farms)

These positive results must not be neglected in any discussion of the difficulties encountered by state firms and cooperatives in hiring out machinery.

A major problem which faces state firms, cooperatives and private firms alike is the low rate of utilization of machinery in the developing countries and the resulting imbalance between costs and revenue. A case in point is Tanzania where tractors were to have been used for field work on 300 hectares per year while the actual performance was only between 25 and 70 hectares[57]. The annual figure for Burma was 70-75 hectares on an average for tractors hired out by state tractor stations.[58]

This low rate of utilization often created a wide gap between the costs incurred for machine employment and the benefits obtained from mechanization.

It would be wrong, however, to conclude that machinery cannot be used economically in the developing countries. What is needed is a thorough analysis of these problems in order to find appropriate answers. Available studies show that the lack of efficiency in the hiring out of machinery by state firms and cooperatives results from a variety of factors in the economic, managing and political spheres.

It can be safely assumed that one reason is to be found in the fact that, in the absence of agrarian policy guidelines, the "business" approach prevails and that managers in charge of hiring out machinery sometimes do not have the motivation which is necessary to promote this new form of mechanization.

In a non-socialist system, the establishment of production facilities run by the state, or by cooperatives, necessarily meets with political opposition from the advocates of a capitalist road of development.

This means that priorities and guidelines must be established in the field of agrarian policy. These aims must be publicized and explained to the population in a convincing manner. The people who are in charge of such programmes must be experts with a political understanding of their job.

A second problem results from the fact that cooperative forms of production were not sufficiently developed to enable machinery to be used efficiently. This led to a situation where tractors and farm machinery were often used on small plots, a procedure which is highly uneconomical. This problem could be resolved by cooperative farms. However, such a development may not be expected over the next few years for most developing countries. This is why semi-collective forms of crop farming are essential. Such forms are now emerging in Tanzania, Somalia and other African countries where they are known as "group farms" or "block farms". We know of specialized production establishments formed by multi-purpose cooperatives in Southern Yemen and Syria, and of integrated crop rotation in Egypt.[59]

Provided they are properly organized, managed and financed, these semi-collective production schemes can make for a much more efficient use of machinery.

A third problem is presented by the fact that technical aspects of the use of machinery have often been neglected. In a number of cases, machinery was found to be ill-suited for the purpose, or equipment had not been sufficiently tested. Some equipment led to uneconomical techniques or was not compatible with other technology. It would seem desirable, therefore, to test any machinery from foreign or domestic manufacturers for the proposed uses and to develop production techniques which suit the conditions in the specific country.

One example that comes to mind is the El Gazair testing station for agricultural machinery which has been operated by the Egyptian Organization for Land Reclamation and the G.D.R. association of farm machinery manufacturers since 1967. The work of this establishment has greatly contributed toward the efficient utilization of tractors and machinery on reclaimed land in Egypt.

A lack of planning has in many cases been found to be responsible for the absence of certain equipment. In most cases, technology is limited to plows, disc harrows and tractors and trailers. This necessarily reduces the rate of utilization of tractors and human labours and revenue and production increases remain low.

It is therefore desirable to extend the scope of operations beyond tilling proper and add one or more of the following operations:

- Land improvement and land reclamation
- Plant protection
- Mechanizing irrigation (for example, by using mobile pump units) and maintaining irrigation works
- Harvesting and threshing

A fourth problem is the general lack of planning and management in the employment of machinery which is often characteristic of developing countries.

Our investigations showed that there were no consistent teams of tractor drivers and repair men. Foremen were little aware of their responsibilities, there was no seasonal planning and little supervision of daily operations.

As a result of this, operations throughout the crop year were insufficiently prepared, work in the fields was done carelessly, vehicles were not used to capacity, there was working on the side, accounting for fuel consumption was neglected, machinery was not properly maintained and field work neglected in favour of transport and handling operations.

This unsatisfactory situation can be remedied by establishing clearly defined teams of tractor drivers and repair mechanics, making foremen fully responsible for the performance of their production team, introducing seasonal planning and supervising daily operations.

The fifth problem results from inadequate forms of employment, remuneration and training for tractor drivers and mechanics, and for foremen and engineers. Our experience is that there is not so much an absolute shortage of tractor drivers and mechanics but a lack of training and motivation. In many cases, tractor drivers and mechanics are employed on a seasonal basis, sometimes as a kind of "day labourers". They consequently regard their work as an odd job and are not interested in using machinery to its capacity or performing operations in good quality. They do not see the need for maintenance and for getting better training. This tendency is often reinforced by inadequate forms of remuneration. Most tractor drivers, mechanics and even managerial staff get fixed salaries or daily wages, and no bonuses are paid for efficient machine utilization, work done in good quality, machine maintenance and shift work. Better efficiency of machine utilization in the developing countries will therefore depend on the introduction of salaries and wages related to performance, the permanent employment of tractor drivers and mechanics and better training for the latter and for foremen.

A sixth factor which is seriously impairing the efficient use of machinery concentrated in state firms or cooperatives in the developing countries, is the lack of maintenance and repair services and the shortage of spare parts. Available investigations indicate that it is not uncommon that 50 % of the tractors and machines are defective at a time. The service life of this equipment is in most cases limited to a few years.

It is therefore essential, in view of the rugged conditions resulting from the soils and climate of tropical countries, to have adequate workshops, maintenance guidelines, repair procedures and spare parts stocks. Training must also be provided for repair personnel.

Another aspect which affects the efficiency of machine utilization is the fact that charges for the services of machine hire stations are made on a different basis from country to country. In some countries, the area worked is taken as a basis, in others there are hourly charges or charges on the basis of fuel consumption. Our experience ist that charges on an area basis (per hectare or per metric ton/mile) provide a good method of calculating the wages of tractor drivers in relation to performance and quality. At the same time, this encourages farmers to make the necessary preparations for a given operation and to give whatever assistance they can.

As regards the amount of charges to be paid, the actual cost of operating the machine must of course be taken as a basis in view of the fact that subsidies will not be available except in a few cases. On the other hand, care must be taken to avoid excessive charges which would make the hiring of machinery uneconomical for the farmer. Charges can be reduced by minimizing the operating cost per tractor or hectare. Preferential rates should be considered for small farmers and for newly formed cooperative farms, a practice which was common for some time in the G.D.R.

Another important factor increasing the efficiancy of mechanization in the developing countries is the gradual establishment of own farm machinary industries. It will be appreciated that mechanization projects relying exclusively in imported technology, involve high costs and potential problems. It is important zu realize, however, that building an entirely new branch of industry calls for large capital expenditures, detailed and long-term planning and a realistic assessment of production and sales capacities. The erection of assembly plants, the acquisition of licenses, scientific and technical cooperation and a possible division of labour between the developing countries and the socialist countries - which have years of experiance in that field - may alleriate the difficult way of establishing an own farm machinery industry in the countries of Asia, Africa and Latin-America.

Throughout its history, mankind has sought more efficient instruments of production. The stone ax and the digging stick, the hand hoe and the wooden plow, the draft animal and the tractor, the combine harvester and the industry-type farms for plant and animal production are characteristic not only of stages achieved in the course of technological development but also of social evolution.

In the words of Karl Marx, it is not important for distinguishing economic epochs what is done but how it is done and what instruments of production are used. These instruments of production are an indicator not only of the development of human labour but also the social conditions in which work is done.

Today, at a time when the countries of Asia, Africa and Latin America are rising from centuries of oppression and underdevelopment, we consider it one of our noblest commitments to clear the way for replacing backward instruments of labour with effecient means of production which will eventually mean more food, less drudgery and a life worth living for the peasants and farmworkers in the developing countries.

References

1 Denker, C.H., Grenzen der Mechanisierung in der Landwirtschaft, Köln 1965, p. 22 and 29.

2 Gallwitz, K., Landtechnik in afrikanischen Ländern, Afrika heute, Köln 1963.

3 Schaefer-Kehnert, W., Handbuch der Landwirtschaft und Ernährung in den Entwicklungsländern, Bd. 1, Die Rolle der Technik in der landwirtschaftlichen Entwicklung, Stuttgart 1967, p. 209.

4 Bean, K.W., Evolution not Revolution in Future Mechanization, World Crops, London, Vol. 15, Dec. 1963, p. 403.

5 Priebe, H., Die Agrarprobleme der Entwicklungsländer im Spiegel der europäischen Entwicklung, Offene Welt, Köln, Nr. 60/1969, p. 102.

6 Smerdon, E.T., Some critical steps in achieving agricultural mechanization in developing countries, Agric. mechanization in Asia, Tokyo, Vol. III, 1971, p. 53.

7 Priebe, H., a.a.O. p. 105.

8 Bean, K.W., a.a.O. p. 405.

9 FAO, Provisional indicative world plan for agricultural development, (IWP) Vol. 1, Rome 1970, p. 231.

10 FAO, without two-wheel tractors.

11 FAO, IWP, as mentioned above, p. 229.

12 FAO, Production Yearbook 1972, Rome 1973, p. 237.

13 FAO, Agricultural Production in Developing Countries in Relation to the Targets for the Second United Nations Development Decade, Monthly Bulletin of Agric. Economics and Statistics, Rome, Vol. 22, 1974, p. 13.

14 Govt. of India, India 1965, A Reference Annual, Delhi 1965, p. 260.

15 Jain, B.K.S., as mentioned above, p. 97.

16 cf.Voss, C., Agricultural Mechanization, Production and Employment, FAO, Monthly Bulletin of Agricultural Economics and Statistics, Rome, Vol. 23, 1974, 1, p. 3.

17 FAO, Agricultural Production in Developing Countries, as mentioned above, p. 2.

18 cf. Lenin, W.I., Der Kapitalismus in der Landwirtschaft, Werke Bd. 4, Berlin 1968, Die Agrarfrage und die Marxkritiker, Werke Bd. 5, Berlin 1966, p. 97, Das kapitalistische System der modernen Landwirtschaft, Werke Bd. 16, Berlin 1971, p. 431, Neue Daten über Entwicklungsgesetze des Kapitalismus in der Landwirtschaft, Werke Bd. 22, Berlin 1971, p. 1.

19 Marx, K., Das Kapital, MEW Vol. 23, Berlin 1970, p. 528.

20 Marx, K., op. cit., p. 790.

21 cf. Soziale Prozesse in der kapitalistischen Landwirtschaft, IPW Forschungshefte, Berlin 1974, H. 1.

22 Griffin, K., The green revolution: an economic analysis, UNRISO, Geneva 1972, p. 70.

23 cf. Michalski, K.-J., Green Revolution in India, Mainstream, New Delhi, Vol. VIII, No. 45, 1970, p. 17.
Michalski, K.-J., Pfeiffer, A., Betriebswirtschaftliche Probleme des nichtkapitalistischen Agraraufbaus in den Entwicklungsländern, Protokoll der wiss. Tagung des Institutes f. trop. Landw. u. Veterinärmed., Leipzig 1970, p. 27.

24 Marx, K., Das Kapital, Vol. I, op. cit., p. 528.

25 Lenin, W.I., Bericht über die Arbeit auf dem Lande, VIII. Parteitag der KPR (B), Werke Bd. 29, Berlin 1970, p. 200.

26 Iswestija, No. 47, Moscow, 6.4.1921 (russ.)

27 cf. FAO, Production Yearbook 1972, Rome 1973.

28 The volume of production of the U.S.S.R. in 1973 lay at 222.5 million tons of wheat, 7.66 million tons of raw cotton and 9.6 million tons of sugar.

29 In the course of the 1971-75, 128.6 billion Rubels will be invested in the Soviet state farms and cooperatives. This is two times as much as that invested in both 5-year-plans 1961-70, cf. XXIV Party Congress of the CP-USSR, Berlin 1973, p. 79.

30 RGW, Statistisches Jahrbuch der Mitgliedsländer des RGW, Moskau 1973, p. 243 (russ.)

31 cf. Roussinow, S., Economic Development of Bulgaria after the Second World War, Sofia 1970, p. 87.

32 Ganew, A., Agrar-Industriekomplexe - eine Form zur Vertiefung der Spezialisierung und Konzentration der landwirtsch. Produktion in der VR Bulgarien, Vortrag, Konferenz d. Verbandes d. wiss.-tech. Spezialisten, Plovdiv, 29.9.72.

33 Schtochov, D., Zolov, J., Organisation und Leitung von Agrar-Industrie-Komplexen, Internat. Zeitschr. d. Landwirtschaft, Berlin/ Moskau 1973, Nr. 6, p. 632.

34 Statistical Yearbook of the GDR 1974, Berlin 1975, p. 187.

35 Kuhrig, H., Zur Entwicklung der sozialistischen Landwirtschaft der DDR, Int. Zeitschrift d. Landwirtschaft, Berlin/Moscow 1974, No. 1, p. 30.

36 Grüneberg, G., Die gegenwärtigen Aufgaben bei der Verwirklichung der vom VIII. Parteitag beschlossenen Agrarpolitik der SED, Berlin 1974, p. 45.

37 Eppner, W., Zur Lösung der Agrarfrage in Kuba unter besonderer Berücksichtigung des bäuerlichen Sektors, Thesis, Leipzig 1973, p. 125.

38 Castro, F., Speach on the 15th anniversary of the land reform, Granma, Havana, Vol. IX, No. 21 (26 May, 1974) p. 3.

39 Anonymous, El desarrollo del plan arocero, Economia y desarrollo, Havana 1970, No. 3, p. 156.

40 Huß, W.; Heynoldt, J.; Wollner, H., Ökonomisch-technische Aspekte der Mechanisierung der landwirtschaftlichen Produktion in Kuba, Report, Leipzig 1971, p. 194 (not published).

41 Ibid. p. 196.

42 Harvesters were used on 7 % of the sugar cane area in 1972.

43 Ibid. p. 54.

44 Ibid. p. 96.

45 cf. Michalski, K.-J., Landwirtschaftliche Genossenschaften in afro-asiatischen Entwicklungsländern, Berlin 1973, p. 271.

46 cf. Voss, C.; Agricultural Mechanization, Production and Employment, as mentioned above, p. 2.

47 cf. Khalatbari, P., Ökonomische Unterentwicklung - Mechanismus - Probleme - Ausweg, Berlin 1971, p. 194.
same author, Überbevölkerung in den Entwicklungsländern, Berlin 1968.

48 cf. Voss, C., Agricultural Mechanization, Production and Employment, as mentioned above, p. 2.

49 Khalatbari, P., Ökonomische Unterentwicklung, Berlin 1971, p. 266.

50 cf. - FAO, Provisional Indicative World Plan for Agric. Development, Summary and Main Conclusions, as mentioned above, p. 24.
- Voss, C., Agric. Mechanization, Production and Employment, as mentioned above, p. 1.
- Abercrombie, K. C., Agric. Mechanization and Employment in Latin America, Intern. Labour Review, Geneva, Vol. 106, 1972, p. 40.

51 Govt. of Tanzania, Second Five Year Plan for Economic and Social Development, Dar es Salaam, 1969-1974, Vol. I, p. 37.

52 Kleinsorge, J. u. Schilling, H., Veränderungen in der sozialökonomischen Struktur der Bauernschaft in der BRD, in: Soziale Prozesse in der kapitalistischen Landwirtschaft, as mentioned above, p. 85.

53 Ibid., p. 102.

54 Ibid., p. 90.

55 IPW Berichte, Berlin 1974, No. 5, p. 75.

56 cf. Michalski, K.-J., Probleme des genossenschaftlichen Traktoreneinsatzes in Tansania, Beiträge zur trop. Landwirtschaft u. Veterinärmedizin, Vol. 7, Leipzig, No. 2/1969, p. 53.
Same author, Grundzüge der agrarwirtschaftl. Entw. Burmas nach der Revolution von 1962, Report, Leipzig 1970 (not published).
Pfeiffer, A., Erfahrungen und Möglichkeiten des überbetrieblichen Traktoren- und Maschineneinsatzes in den Entwicklungsländern, Report, Leipzig 1971, (not published).

57 Michalski, K.-J., Probleme des genossenschaftlichen Traktoreneinsatzes..., as mentioned above, p. 100.

58 Michalski, K.-J., Grundzüge der agrarwirtschaftlichen Entwicklung Burmas..., as mentioned above, p. 50.

59 cf. Michalski, K.-J., Landwirtschaftliche Genossenschaften in afro-asiatischen Entwicklungsländern, Berlin 1973, p. 335.

Social Aspects of Farm Mechanization in India

by Z. A. Ahmed

Allow me to point out that I fully agree with the analysis made by Prof. Michalski in the note he submitted to the conference on the question of mechanization of agricultural processes in underdeveloped countries. I think this analysis is quite right. As for the conclusions which he has drawn from the analysis, I think - quite frankly speaking - they are also correct. Therefore, I would like to congratulate him on something which - in our opinion will give a proper understanding of this question in the underdeveloped countries.

In India, there have been various schools of thought on the question of mechanization of agriculture. There has been one school of thought represented by people belonging to the Ghandian ideology, the ideology of Mahatma Ghandi. They have opposed mechanization. One of there main objections was that machinery will replace man and create unemployment.

And they have always said that in a country where the man - land ratio is characterized by too many men living on a limited area of land and by an increasing number of men drawing their sustenance from agriculture machinery will cost a lot, and lead to hunger, unemployment, more poverty. The All India Kisan Sabha, which is the main organization of the Indian peasantry - a very old organization -, has opposed this point of view. The All India Kisan Sabha representing millions and millions of peasants does not agree with the idea of "preventing machinery from being introduced because of the increasing poverty and unemployment resulting from it". In fact, there is a tremendous amount of unemployment in the rural areas without the machinery being there. The reason for this - as Prof. Michalski has very correctly stated can be seen in the underdevelopment of both the whole agriculture and the whole economy. It is quite obvious that this problem of unemployment and poverty cannot be solved without raising the level of production in agriculture. If you do not raise the level of production, population will go on growing, land holdings will become smaller and smaller - as you know, we are living in a capitalist society - and the division of those holdings will go on. Thus, the number of uneconomic holdings will be growing and agricultural production processes or agricultural enterprises will become more and more uneconomic in these very small holdings. So, you cannot attribute unemployment to the use or non-use of machinery. There are some other, basic causes of unemployment in India. That is why we, that means the All India Kisan Sabha, have opposed this point of view. This point of view seeks to keep Indian agriculture - the technical base of Indian agriculture - very backward, stagnant and underdeveloped. There won't be any promising future for Indian agriculture if you stay where you are.

There is another point of view represented by the modern bourgeois economists, who have a capitalist outlook. They want the mechanization of agriculture but, according to them, machinery should be used for the development of a class of substantial capitalist farmers. They expect those capitalist farmers to develop the agricultural resources of the country and to constitute the basis of a reorganised agricultural society.

While supporting the idea of using machinery in agriculture, we have always opposed the idea of using this machinery for creating a class of parasitic landholders. There is no doubt that they would use these modern machines for developing some amount of production, but they would also deprive the mass of the small and middle peasants of the advantages that could accrue to them as a result of the use of that modern machinery. This class of parasitic landholders would monopolize all the advantages of the modern machines. The important point to be noted is that the Indian government, which is not a socialist one, has accepted this conception of mechanization, which means mechanization for a small class of big farmers who are given subsidies and who are provided with all the necessary facilities. They get millions and millions of rupies and this is done in the name of an "agricultural development". This has been the policy of the government of India. Have a look at the four Five-Year-Plans and you will find that they have provided enough funds for the development of a class of capitalist farmers - big peasants - who, in their opinion, are the basis of the "green revolution" in India.

There is a lot of talk about the "green revolution" in India. Some people think that the "green revolution" has been completed in India. They already want the "white revolution", which is the "milk revolution". But we have taken up a clear position concerning this question. All the money that has been spent by the government of India and all the mechanization that has taken place as a result of the efforts made by the Indian government and as a result of the initiative taken by the bigger landwoners themselves have contributed to the creation of a small class of farmers who have undoubtedly increased production to some extent. We cannot deny this fact. They have increased production, but this type of a "green revolution" cannot become the basis for transforming Indian agriculture and for raising Indian agriculture technically, materially, from the production point of view and from the social point of view to a higher level of organization and production. The limitations of the "green revolution" that has taken place in India must be seen in close connection with the limited mechanization in Indian agriculture.

I would like to mention another important fact. There is a lot of talk about the "green revolution" but the maximum amount of arable land that has come into the orbit of the "green revolution" is 15 % and 85 % of the land still remains outside. On 85 % of the land the use of machinery is almost nil and the use of fertilizers is also very low. There is also a very poor irrigation - 85 % of the area are non-irrigated. But, as you all know, the question of irrigation is the biggest problem in tropical and subtropical countries. The use of fertilizers depends on the irrigation and even the use of machinery is limited by the possibilities of irrigation. I would say 92 - 95 % of the peasants are without proper inputs for the development, including machinery. The "green revolution" - a revolution based on the development of a class of capitalist farmers - has not laid the foundations of a new agrarian society. A different path will have to be found. In this connection we are putting forward the idea of a non-capitalist path of development in agriculture. Up to this time, the government and the ruling classes have followed the capitalist path of development.

There are the penetration of commercial capital into agriculture, the development or the influence of monopoly capital in the agricultural market and the looting carried out by the mechanizm of prices. Furthermore, I would like to mention the domination of the commercial capital over the agrarian market and the increasing production of commodities. In addition to this, there is the development of a limited mechanization. I would speak of a limited mechanization, although, India has - compared to other Asian or African countries - more tractors, factories etc. But even this does not improve the situation in Indian agriculture because of these machines being confined to a narrow stratum of the agricultural society. Therefore, we have to develop and to work out the conception of a non-capitalist path of development in agriculture. This is, I think, very essential!

The All India Kisan Sabha has tried to lay down some sort of an outline for the non-capitalist path of development. What are the features of the non-capitalist path of development? I will just briefly mention those here. I would like to read out a few paragraphs from my report. The "green revolution" is based on an agro-technical revolution in the process of cultivation. But this is not the essential point. The "green revolution" is mainly based on the utilization of a limited amount of machinery by a narrow stratum of big farmers who depend for increasing production not so much on the use of machinery - this is a peculiar feature of India - but on the use of high-yielding varieties of seeds. I think this is a very interesting phenomenon. It is not the agrotechnical base that has developed on a very big scale but it is the use of high-yielding varieties. I would like to mention some Mexican wheat - the dwarf wheat of Mexico. We introduced this wheat after 1966, as you all know, and suddenly more tractors and more machines are being used. The introduction of this Mexican variety of wheat has stimulated it. The limitations of the "green revolution" can also be illustrated by the following fact. There was a great noise made by certain circles of the Indian government about the "miracles" of the "green revolution", but, nevertheless, the general rate of agricultural development is very low. We talk so much about "green revolution" and mechanization but the general rate of development is very low. As for the period from 1952 to 1970 the annual growth rate in agriculture - referring to the country as a whole - was 3.1 %. The states in which the growth rate was very low are:

Bihar	0.7 %	Madhya Pradesh	1.5 %
Assam	1.4 %	Uttar Pradesh	2.5 %
West-Bengal	1.5 %	Rajasthan	2.6 %

I do not want to go into details with this. Nevertheless, you must understand that this is all what the "green revolution" has brought about. The reason for this is the non-participation of the mass of the peasantry. In the process of the "green revolution" the growth rates come to 1 %, 1.5 % or 2 % while the population is growing by 2.5 % to 3 %. So, people are confronted with a permanent intensification of their exploitation and on ever increasing poverty. Another important fact is that the "green revolution" has accentuated rural inequalities. It has not led to a general raising of the limited mechanization in high-yielding varieties but it has led to a growing inequality - the big landowner is getting richer and the small landowner is getting poorer. What we are fighting for is not the abolition of the use of machinery. On the contrary, we want more and more utilization of machinery serving the mass of the poor and middle peasants. This is the main thing! And therefore, basic institutional changes must be made. In the face of these socio-economic relations in agriculture, the conditions of production will have to be transformed. It is quite clear, that if you maintain this whole set-up, you will not be able to attain the results wanted.

I myself, I am a middle farmer and therefore, I know the problems. You go to a bank and you do not get any money. You want to buy a tractor - but it costs between Rs. 25,000 and Rs. 30,000 on the "black market" it costs even more. How could I - as a man owning 7 hectares - get Rs. 30.000 for buying a tractor?

As you all know, we have a certain ceiling of 7 to 10 hectares. But there is the big farmer who has 200 or even 500 hectares. He has concealed all this land and tries to evade the ceiling. This ceiling is nothing else than a sheet of paper because everybody has parceled out his land among his family. He has divided up all the land and if you have a look at the papers, you will not find any surplus land with him. So, our ceiling is fictitious! As for our cooperative societies, they only give credits to the rich peseants, the big farmers. The poor man does not get any credit at all.

The question of irrigation facilities and fertilizers has become very acute. We do not have enough fertilizers and there is a permanent rise in prices - prices have even doubled!

More than 80 % of the Indian peasants have less than 2 hectares and 60 % to 70 % of the peasants have less than 1 hectare. If one of those peasants owning about 1 hectare asks for a credit, he will not get anything. From where and how could he get a credit? Therefore, the non-capitalist path of development has to be adopted - the whole path of development has to be changed.

The use of machinery is not only a question of technology and quantity. The problem cannot be solved by producing more tractors etc. In fact, we have come to a dead end of the road! There is only one possibility for our agriculture - the transformation of the socio-economic relations and the relations of production. Otherwise, Indian agriculture cannot advance. Therefore, we have put down the main aspects of the non-capitalist path of development. What does the non-capitalist path of development imply? In terms of institutional changes and changes in the socio-economic relations I would like to mention the following features.

(1) The non-capitalist road of development, first of all, implies the complete liquidation of all the remnants of feudal and semifeudal exploitation. There are a lot of old feudal remnants or feudal and semifeudal forms of exploitation in India. Think of the share cropping and the tenant system. Above

all, these forms should be completely eliminated. This is one of our most important tasks. We have to abolish all the dust and dirt which we have inherited from the past.

(2) The non-capitalist road of development also implies the breaking up of the concentration of land in the hands of the rural exploiters. There is a high degree of concentration. The old feudal landlords were allowed to retain a lot of land - they grab a lot of land. Land grabbing is going on! The poor peasant is forced to sell his land because he has no bullock and the money lender wants to be paid. So the Harijans surrounding any plot of land came to me as a small farmer and said: "Buy our land. It costs Rs. 1,500 a hectare". They sell their land below the reasonable price and it is the big man who is quietly buying this land.

Therefore, we have to break up both the feudal and the capitalist concentration of land in the hands of the rural exploiters. We also need real ceilings for the land holdings. They must be effective and not fictitious! Obviously, we cannot have very high ceilings. In the days of your land reform in the G.D.R. the ceiling was 100 hectares. This would be absolutely impossible in India. Our ceiling law speaks of 7 hectares (good quality of land) to 15 hectares (very poor quality of land). Even this is quite high. But, as I have already said, it is not effective.

(3) The non-capitalist road of development also implies the adoption of effective measures for curbing and preventing the further growth of capitalist forms of agriculture. The positive alternative of this is the development of cooperative forms of association of the poor and middle peasants - cooperatives of various types.

What we need in the first stage are service cooperatives in a large scale. We need cooperatives for irrigation, cooperatives for marketing, cooperatives for getting and for using fertilizers and also cooperatives for such processes as cultivation. I am not jumping to the stage of cooperative farming. Cooperative farming is a feature of a higher stage. We have seen it in the G.D.R.

But the form of association has to be developed and if we proceed along that line, it will be possible to curb the economic growth of the individual big capitalist farmers.

(4) The non-capitalist path in agriculture implies plans, projects and programmes which must be applied to the whole country and which are to promote the initiative and capacity of the entire mass of tillers of the soil.

I am speaking of joint programmes and joint projects which could develop the production initiative and capacity of the mass of the tillers. This is the only way for attaining higher results in production.

(5) The non-capitalist road also implies a determined and radical transformation of the structure of the market in agricultural production. The market has to be controlled. We cannot leave it in the hands of the monopoly capital.

We think it is quite possible - in a country like India where the peasant movement is growing - to have a more determined and broad-based organization of the peasantry, which means the unity of all democratic forces behind one agrarian programme. This is coming into existence.
I am quite convinced that any narrow sectarian approach to this question will be ruinous. It is necessary for us - according to our class point of view and our political point of view - to form a broad, firm and mature movement for implementing our aims. Our main attack has to be directed against the big landlords, the capitalist landlords and the feudal landlords - not against the rich peasants! In our organization there are some friends who also want to fight against the rich peasants. But I say: No! The rich peasants have the status we are not going to touch. The rich peasant is not our enemy although he is an exploiter of the labour of poor peasants, middle peasants and agricultural workers.

There are so many parties in India but a broad-based programme should be like this: There should be more than the support by the Communist Party or this single party or that single party only. We need the support of various other parties, including the ruling sections of the governing Congress Party. The Congress Party comprises some different political groups. There are the centre, the groups right to the centre, the groups left to the centre etc. They are behind an agrarian programme. A powerful mass movement based on the unity of all democratic forces as well as mass struggles and mass initiatives will make it possible to create the conditions for the introduction of the non-capitalist road of development in agriculture. Such a non-capitalist path of development would not only raise our agricultural production from the technological point of view but it would also set into motion such forces, elements and correlations of class forces that will bring about far-reaching changes in our economic and political structure. This refers to the country as a whole.

Recent Stage and Development of the Utilization of Farm Machinery in the Arab Republic of Syria

by Dr. Al Jumaa

The tempo and level of agricultural development of a country are to a large extent determined by the growth of its agricultural production and the rise in the productivity of labour. In this respect significant value is attached to the gradual mechanization of the agricultural activities both in the plant and livestock production.

We have formulated our policies and decisions with regard to the projected agricultural development in the Arab Republic of Syria on the basis of these facts. Today we are of the opinion that it is necessary in the developing countries to replace gradually and systematically the traditional production methods and the vestiges of pre-capitalist patterns of production techniques, by the introduction and development of mechanized production methods. In detail, we have based the necessity of mechanizing the agricultural production in Syria on the following urgent socio-economic and techno-economic view points:

1. The progressive transformation of agriculture in Syria is characterized by the creation and development of state agricultural farms and the establishment of agricultural cooperatives (mainly multipurpose co-operative societies and also a considerable number of co-operative farms). The realization of the advantages of these progressive types of agricultural farms directly depends on the effective use of tractors, farm machinery, power pumps etc., i.e. on the mechanization of agricultural production.

2. The Arab Republic of Syria possesses a vast reserve of agricultural land which can be utilized for the expansion of the cultivation of cereals as well as the production of other agricultural crops. An effective utilization of these resources, however, is possible only by the application of scientific technique.
3. The economic utility of such costly and investment consuming new agricultural land-projects, like the Euphrates Project, can be realized only by means of modern technique and technology. Traditional production methods and primitive techniques, which are still today in practice in many Syrian villages, are absolutely unsuitable for these projects.
4. Moreover, the intensification of agricultural production in the traditional agricultural regions calls for the adoption of modern means of production. A fundamental task in this respect is to introduce tractors and agricultural machinery into these regions. For the successful application of these machines, the prerequisites for the effective utilization of the measures for further intensification must first be satisfied, for example, use of high quality seeds and agrochemical inputs.

The continuing process of the mechanization of our agriculture has, of course, some limitations which result from the social conditions and the limited investment capacity of the national economy of Syria. That is why the mechanization of the working processes has been concentrated first of all on those crops in which the rise in the production and increase of quality are rapid and high, for example wheat and barley.

As a result of these purposeful processes of mechanization which have already begun, about 344 man-hours per hectare could be saved. Due to this mechanization not only a high and profitable harvest has been ensured, but also enough resources have been made available for the further intensification of the agricultural production and the bringing of additional land under cultivation as well as for the development of livestock production.

Lastly, as a result of mechanization the income of the state farms and agricultural co-operatives has increased which has also been reflected in the improvement of the working conditions as well as the living standard of the agricultural labour force employed on those farms.

The experiences of previous years have also revealed that in some cases tremendous difficulties had to be faced with the mechanization of Syrian agriculture. To take an example, until 1964 the import of tractors and other farm machinery in Syria used to be carried out without state control. The businessmen could purchase tractors from all over the world. As a result there are 39 types of tractors being used in the agriculture of Syria today. Besides this, there were no organizations and service stations for testing and examining the foreign tractors and farm machinery with respect to their adaptability to the prevailing conditions in Syria. Under these circumstances nobody can doubt that as a result of the presence of many different types of tractors great difficulties had to be faced for the application of tractors, their service, maintenance repair and reconditioning, supply of spare parts and training of operators for the service of these tractors. Moreover, it can be also pointed out that most of these tractors and farm machines are still under capitalist possession and as a result the small producers remain deprived of farm machinery.

On the other hand, the modern techniques of agriculture which are being increasingly used in Syria will also cause a remarkable development of the productive forces and production relations in agriculture. The mechanization of state farms and agricultural co-operatives has in particular born out this positive development in Syrian agriculture.

Before considering the special aspects of the establishment of a state-owned industry in Syria for the production of farm machinery, and the problems of machine utilization and drawing some conclusions about the future, some remarks must be made first of all about the present level of agricultural mechanization in the production of cereals and cotton as well as in livestock production.

1. Production of Cereals

Wheat and barley are grown on the largest part of the farm land. The production of 1.8 million tons of wheat and about 0.9 million tons of barley has been planned for the year 1974. About 75 % of the cereal growing area is suitable for mechanization. Our potentials and perspectives, however, are still not enough to cultivate and harvest all these agricultural fields completely by machines.

For example, in 1973 about 12,224 tractors were available for field work, by which only 50 % of the cultivable area could be cultivated. The tractor requirement for these cereal growing areas alone is about 14,776.

Besides, we have a relatively large number of combine harvesters. The total number is about 1,400. According to our calculations we "only" need about another 250 combine harvesters to be in a position to harvest the cereal crops within an advantageous period of time viewed from an agronomical standpoint.

2. Production of Cotton

Cotton is an important export product for Syria and at the same time also a decisive raw material for her own industry. The volume of cotton production per year is about 410,000 tons. Many of the operations, especially harvesting, must be carried out manually. One labour force is required to harvest one hectare of cotton. Our studies have revealed that in none of the districts of Syria is the necessary number of labourers available which are required for the cotton harvest, according to the above proportion. This situation necessitates a forcible extension of the normal period for harvesting causing great losses and deterioration in quality effect of the cotton and thus having a negative effect on the national income. Therefore, it is an accute economic necessity to extend the level of mechanization of cotton production to include further operations, especially as regards to harvesting the level of mechanization has up till now been limited mainly to seed bed preparation.

The Syrian government is at present endeavouring to solve this problem by importing cotton harvesting machines from abroad and also by producing its own farm machinery. In this respect greater importance will be attached to the task of impressing the farmers about the advantages of pooling their cotton growing fields together and also about the advantages of establishing an "integrated crop rotation".

The proper control of pests plays an extremely important role in the cultivation of cotton. The state imported 481 sprayers of big and medium size available for this purpose. These sprayers are used mainly on the state and co-operative sector.

Moreover, twelve economical aircrafts are also available for pest control, which are used for spreading fertilizers as well, in addition to the pest control operations.

With respect to the production of sugar-cane, which covers an area of 11,000 hectares, it can be said that for mechanizing the harvesting of sugar cane the similar difficulties have to be overcome as those encountered in the harvesting of cotton.

3. Livestock Production

Animal production is also of great importance in the Arab Republic of Syria. At present the animal husbandry including the feeding of animals as well as the processing of livestock products are still done by traditional methods in which the costs of manual work are high.

In this respect sheep farming must first be mentioned. Syria has a sheep population of about 6 millions, 85 % of which are held by the nomads in the desert and semidesert areas of Syria. The mechanization of the husbandry of the remaining sheep population (15 %) serves the interest of the 78 agricultural co-operatives specialised in sheep farming and the 28 agricultural co-operatives dealing with sheep breeding. Moreover, there exists in each district one centre for sheep farming. In these centres and in the co-operatives there is a big demand for fodder producing machines, means of transportation, processing plants for milk etc. which cannot be met at present.

A similar situation prevails in cattle farming. The state centres also act in this respect as model farms and demonstration farms for the mechanization of milking operation and milk processing. For example, 8 state farms have been equipped with milking machines and watering devices. Some of them have even mechanical devices for clearing dung. On the basis of the approved development plans and the contracts made with foreign companies 26 more state dairy farms with a high degree of mechanization will be established in the future.

For this purpose the simultaneous importation of 6, 000 heifers and cows with the best dairy performance has also been planned.

In addition to the state farms, it will be necessary in future to support also those co-operative societies which deal with livestock production by direct help from the state regarding the mechanization of their production especially milk production. Finally, strong measures are also to be taken for the mechanization of poultry farming. Such measures are necessary mainly for the intensive poultry farming stations and slaughter houses for poultry birds.

It can be seen from what has already been said that remarkable progress has been achieved in the mechanization of agricultural production in Syria inspite of the limitations and constraints which have been already mentioned. Last but not the least to be mentioned about the efforts of the Syrian government, is its plan to develop a national industry for the production of farm machinery and to establish repair work shops for these machinery. For this purpose a state-owned factory has been constructed in Aleppo which already has a branch factory im Homs. The objectives of this factory, which was constructed in 1969, are as follows:

1. To produce tractors of medium capacity,
2. To produce ploughs and other agricultural implements,
3. To produce diesel power pumps with capacities ranging from 15 to 90 H. P.
4. To act as a development centre for the gradually growing branch of the industry which deals with the production of farm machinery and tractors.
5. To train and prepare expert personnel for the production plants of the farm machinery industry which will be established in the future.

In 1972 8 to 10 tractors of type EBRO 160 (Spain) could be produced daily. These tractors are assembled from imported parts on the pattern of a French model. In addition to that, the factory imports at present modern farm machinery, for example combine harvesters and tractors of a higher capacity. These home-produced and imported machines are sold to the co-operative and private sectors against cash payment or against payment on an instalment basis through the agricultural bank at the interest rate of 5.5 % per annum.

In the meanwhile the branch factory in Homs has already undertaken the production of fertilizer spreading machines and ploughs. 5 agricultural implements, mainly share ploughs and disk ploughs, are produced in this factory daily. It is worth mentioning here, that these machines have proved effective in practice. This factory must be expanded in the future by the establishment of three more branch factories in order to meet the increasing needs for the mechanization of agriculture. The development programme has for this purpose the provision of estblishing the following factories:

- a factory for the production of farm machinery
- a factory for the production of small three-wheel tractors with 16 H.P.
- a factory for the production of individual spare parts or construction materials. This factory will serve as the delivering factory for the tractor plant in Aleppo.

It is well recognized today, that the establishment of a state-owned farm machinery industry is of great importance in view of the necessity for increasing agricultural production and its modernization in Syria. Finally, we are of the opinion that to attain these objectives, the available potential and prospects of farm machinery in the private sector must be efficiently utilized under the influence of the state. This relates to the existing small and medium size private factories in Syria which produce seed drills, ploughs and other implements suitable for local conditions.

In the following paragraphs we shall concentrate our discussion on some of the relevant problems concerning the application of farm machinery in Syria. As we know from our experience, it is useless to provide agriculture with tractors and machines unless the construction and development of repair work shops, overhauling centres and the establishment of maintenance and service stations are started at the same time. The carrying out of repair work on tractors and farm machinery has in the past been the monopoly of the private sector, although they were not at all able to meet the total requirement for repairing machines. For this reason, the Syrian government has taken into hand the task of establishing state as well as co-operative repair workshops and service stations.

In the meantime the Syrian Federation of Co-operatives has opened two co-operative repair workshops and service stations with assistance from the Ministry of Agriculture. The experience gained in this respect since 1973 indicate the necessity and usefulness of such workshops and centres and induse us to continue the further establishment of such specialized factories. The operation of mobile repair workshops has also proved very useful in the different districts during the sowing and harvesting seasons. These repair workshops also undertook the sale of spare parts at the same time.

On the basis of the mechanization policy which our government has embarked upon during the last few years, it can justifiably be concluded that as a result of this policy a decisive contribution has been made towards the consolidation of the non-capitalist way of development in Syrian agriculture. As already indicated, we place more importance on co-operatives as regards the application of machinery, because according to our experience they are more suited to the task of carrying out the gradual mechanization of Syrian agriculture and above all they are best suited to the interests of the small farmers.

In view of this, the ruling Baath Party and the government support the development of co-operatives with the following objectives:

- to impress upon the working peasantry the correctness of the co-operative system;
- to establish co-operative democracy in order to raise the level of confidence among the members about their co-operative society;
- to promote and help the process of development of the agricultural co-operatives from lower types to higher forms of co-operative farms; and
- to provide the agricultural co-operatives with tractors and machinery at reasonable prices or against credit under favourable terms and conditions.

The gradual transition to co-operative farms serves the cause of social progress. For this reason we have begun to induce and organize the members of the multipurpose co-operative societies to undertake co-operative farm work in addition to the co-operative purchasing and use of machinery, because the prerequisites for the effictive use of machinery can be only met in this way.

These possibilities were not, in fact, fully utilized in the past. For instance, some members of the co-operative societies purchased tractors and combine harvesters and started to organise their own private machine lending business which was a completely violation of the statutes and directly against the progressive development principles of the Syrian co-operative movement. For this reason suitable measures have been teken by the government to remove such capitalist tendencies including all of the contradictions associated with them. For example, it has been forbidden to sell big farm machines individually to members of agricultural co-operatives. The problem facing mainly the common use of machinery in the co-operatives was at the beginning the limited field of use of the machines. The costs of machine use, which are directly related with the yearly performance of the machines, could often not be covered by the farm receipts.

According to the information available to us, up to the middle of 1973 out 497 tractors 125 (= 25 %) were used with financial loss. That is why a substantial number of co-operative officials and members consider the available technique as an economic burden. Only in very rare cases, however, can the actual causes responsible for the unecenomical application of tractors in a number of co-operatives (this is naturally also true for some of the state farms) be analysed.

According to our opinion the factors responsible for that are as follows:

(1) The sense of co-operation and the co-operative spirit are not adequately developed, which manifest themselves in the indifferent attitudes of the members with regard to the co-operative properties. Very often the managing committees are not in a position to carry out the co-operative norms.

(2) There is a considerable shortage of experienced and qualified tractor drivers, mechanicians and organisers for the utilization of the machines. As a result of inadequate service, maintenance and overhauling facilities a long time is needed for repair and overhauling work.
Secondly, due to the necessity of shifting machines from plot to plot and because of the small parcels of land, the daily performances and brigade performances become lower.

(3) The level of planning, organization and accounting of the use of machines has still not been developed to a sufficient degree. Setting aside the problems concerning the mechanization at the very beginning, we shall devote our attention to the question of how to further develop the co-operative as well as the state organized utilization of machinery giving special consideration to the problems which have been analysed above. In this respect we shall concentrate on the two major points which are as follows:

1. Establishment of production and well mechanized demonstration and model farms under state control for the production of primary agricultural products as well as for the finishing and processing of agricultural commodities.
2. Gradual establishment and development of efficient co-operative production farms on the basis of solid arrangements with modern agricultural techniques.

In this respect it is necessary to eliminate the shortage of expert personnel and to make conscious efforts for the establishment of a state education and training centre for training the planners, organisers, tractor operators, mechanicians and other experts in service and repair works. It is also further necessary to reduce in future the number of types of imported tractors and machines, so that from the very beginning the servicing and repairing of machines including the supply of spare parts do not become unnecessarily difficult.

Finally, it is most important to gradually overcome the problem of the fragmentation of land holdings by the establishment of co-operative farms in order to prepare suitable conditions for the application of modern agricultural techniques. For all of these major tasks mentioned above it is necessary for the Syrian government to further control and also financially support the process of gradual mechanization of the agricultural economy.

In conclusion, it can be said that the results which we have so far achieved seem to us very hopeful. On the other hand, we can also utilize in future the experience of other friendly countries as regards the mechanization of agricultural production. We shall be able to provide the increasing population of the Arab Republic of Syria with food and nutrition and industry with agricultural raw materials and also to increase the volume of our agricultural exports only by means of the co-operation among the state and co-operative sectors and the progressive individual farmers.

The Level of Development and Future Outlook of Agricultural Mechanization in Cuba

by E. P. Grannum Bonora

In this paper we want to give an introduction to some basic questions about the progress of agricultural mechanization in Cuba in the past and present and also the future outlook for mechanization in the years to come.

Due to her natural conditions Cuba has always been a land with an important agricultural and livestock economy. The predominance of feudal production relations, before the victory of our revolution, caused an extensive but irrational use of land. As a result of the limitations arising from the prevailing mono-culture many resources had not been utilised which could otherwise have played a significant role in the national economy.

In the past about 1700 tractors were imported annually. According to FAO estimates Cuba had 9211 tractors available in 1960, i.e. 258.8 hectares of arable land per tractor. Fertilizing which is one of the most essential requirements for the cultivation of sugarcane, was completely lacking. In addition, there was a lack of a proper infra-structure, irrigation facilities and field paths. During the harvest of sugarcane each single cane had to be cut and loaded by hand. It was the same method which was in practice during the colonial period. Only a few financially strong organizations could carry out some particular agricultural operations with the help of machines. But the ordinary peasants had no other choice but to use animals as the only source of tractive power and on the top of it there was a considerable shortage of agricultural implements.

Many great obstacles had to be removed in order to overcome this situation: The inadequacy of infra-structure, the totally inadequate standard of mechanization and the lack of a qualified labour force associated with it.

The importation of tractors was increased during the three years between 1959 and 1961 to 11,621, i.e. three times as much as in the past five years. As a result of this policy the amount of farm land per tractor in the state-owned sector has been reduced to about 80 hectares in 1968.

It is well known that the mechanization of agriculture does nót only involve the use of tractors. That is why efforts were made to provide at the same time a large number of machines and implements for various other purposes.

In order to give you an idea of the importance of agricultural mechanization in our country it is sufficient to say that our revolutionary government put aside the highest investment sum in the national economic plan for this purpose. By means of mechanization, our country will finally overcome the problem of labour shortage, thus the fundamental basis for the possibilities of increasing production and reducing costs per unit of production will be provided.

The Potentials of Mechanization and the Organizational Aspects of Using Machinery

In accordance with the agricultural development plans the potential for the mechanization of agriculture has increased enormously. In 1958 about 8,000 tractors were in use in Cuba.

In the last ten years of revolutionary development the number of tractors has increased and the number rose to 37,807 in the year 1969. This means that for every 1,000 hectares of cultivable land a total of 13.1 tractors are available. The number of tractors has already increased to 51,600 in 1970. This raises the number of tractors available for every 1,000 hectares of cultivable land to 16.3.

This quantitative analysis is not, however, enough to present accurately the actual situation. The rapid development of agricultural mechanization in quantitative terms bears enough evidence of the considerable efforts of our revolutionary forces. It was possible due to the unselfish help of Soviet Union and other socialist countries. But this rapid mechanization did not correspond to a simultaneous increase in the productivity. This lag indicates without doubt the absence of a rigorous and rapid process of political, economic and social transformation during the course of the revolution. This also reveals the organizational weakness during the first few years of the revolution which stems from the limited number of qualified experts.

Considering that the process of mechanization must bring about an increase in the productivity of labour, extra-ordinary endeavours must be undertaken by the leading forces of our revolution to organize and develop agricultural technique as well as to make full use of it. The process of organizing agricultural technique including the introduction of work norms was carried out jointly by the Ministry of Agriculture and the Ministry of Labour. For this purpose a pilot scheme was first carried out in the province of Matanzas.

The following achievements in the cultivation of the farm land in the first half of the year 1972 can be cited as concrete results of the mechanization:

- 38 % increase in the productivity of labour,
- 23 % rise in the production,
- 18 % decrease in the fuel consumption,
- saving of 243 labourers, and
- economization of the operational time of 8, 242 machines and implements.

This plan of organizing and norming work units in agriculture was carried out throughout the whole country and positive results have already been reported from all parts, although the execution of the plan is not yet complete.

The Mechanization of the Most Important Agricultural Crops

One of the limiting factors for increasing agricultural production is the shortage of labour, the solution for which requires the raising of the level mechanization in the cultivation of the main products. Table 1 reveals the present and the planned level of mechanization of the main crops. We can see from this table that it will be possible to save about 50 % of the labour costs by mechanization.

According to the future plan up until the year 1980, the major emphasis as regards the employment of labour force, will be placed on the cultivation of sugarcane, citrus fruits and tobacco.

Progress in the Mechanization of Sugarcane Cultivation

The cultivation of sugarcane underwent a deep and all-round transformation involving all the phases of its production. As a result of this transformation the traditional system of cultivation was completely abandoned.

The vast area of land which is required by sugarcane cultivation, the low degree of mechanization at the beginning of the transformation and the enormous labour force needed due to the low level of mechanization were the main reasons behind the efforts to improve production techniques. At present approximately 90 % of the cultivation and fertilising work has been mechanized.

Herbicides are used in advance for controlling weeds. Use of aircraft, in the same manner as their use in applying nitrogenous fertilisers, plays a leading role in this respect. At first Cuban and Soviet cutter-machines were introduced for harvesting sugarcane. These machines were later developed into fullfledged harvesters which operate very successfully. Moreover, of 89 % of the sugarcane which are at present harvested by hand, about 90 % are loaded mechanically. The progress of mechanization of sugarcane harvesting in different harvesting periods has been reflected in Table 2.

At present about 60 % of the harvested products are transported by trailers of tractors, the rest is transported by lorries or cars drawn by animals.

The efforts which were made to raise the level of mechanization in sugarcane cultivation indicate that in the future it will be possible to mechanize all the phases of production, although at present we are just in the beginning of this process.

Some Aspects of Mechanizing Rice Cultivation

During the pre-revolution period only 1.4 % of the producers, who accounted for 70 % of the national production, had a relatively high degree of mechanization. The rest of the producers used age old agricultural techniques without absolutely any mechanization.

The revolution, which started with the execution of an agricultural reform, brought about considerable increase in the rice production. The following factors were the reasons behind this:

- changes in the hitherto existing technology,
- introduction of machinery in place of manual work,
- use of the new rice growing areas,
- great progress in the breeding research for developing new rice varieties.

The lack of harmony in the construction of irrigation systems and levelling of new cultivable land, poor organization of production as well as lack of qualified people, led to the fact that in spite of a high degree of mechanization (97 %) the increase in the production during the first few years resulted mainly from the increase in the areas under cultivation.

But we can point out that since 1971 a positive turning point has ben reached. The increase in the production since that year can be attributed to an increase in the yields per hectare. The yields rose to 1.59 tons per hectare in 1971 and 2.20 tons per hectare in 1972 as against only 1.46 tons per hectare in 1970. At present experiments are under way to replace aircraft by agricultural machinery for sowing seeds and spreading fertilizers and also to increase the performance of the combine harvesters.

Some Aspects of Mechanizing the Cultivation of Citrus Fruits

The cultivation of citrus fruits in the pre-revolution period was carried out mainly on the basis of manual work. At that time the citrus fruit growing area was only 13,500 hectares. However, already by the year 1973 the cultivation of citrus fruits covered more than 60,000 hectares.

In the projected plans up until 1980 provisions have been made to expand the citrus fruit growing area to 270,000 hectares. Considering this planned expansion, experiments are being carried out at present to improve the cultivation techniques including the introduction of newly developed plant varieties.

The results of these activities manifest themselves in the transformation of the technological process of the cultivation of this crop specially in mechanizing the following operations:

- Filling plastic bags with the bulk of the organic substance and the distribution of bags of seeds in the nursery grounds
- Digging pits for the seedlings and the use of fertilizers during transplantation by mechanical devices
- Mechanical and chemical devices for weed control
- Use of plant protection materials
- Mechanization of the irrigation system
- Mechanization of hatch reaping
- Partly mechanized harvesting
- Transport in containers

The majority of these experiments and their findings have been put into practice. Some of the findings are still under further trial.

Progress in the Mechanization of the Cultivation of Vegetables, Root and Tuber Crops

During the pre-revolution period production of vegetables was extremely low. At present, however, the production has increased considerably. Previously the production was done mainly by manual work. Today the majority of the operations, especially in harvesting, which accounts for the major portion of the cost of manual work (50-60 % of the total cost), have been mechanized; although the degree of mechanization is still very low (15.2 %). At present experiments are being carried out to obtain only those plant varieties which are suitable for mechanization. We have achieved success in the mechanization of the seed beds and in uprooting seedlings as well as in direct sowing of seeds after necessary dressings.

Although in the experimental phase, the latter device will facilitate the direct sowing of seeds and will help in economizing the application of seeds. The majority of the operations are already carried out mechanically. This is particularly true of fertilizing, the use of herbicides and the control of plant diseases and pests.

The cultivation of root and tuber crops ensures considerable expense due to manual work especially in harvesting, although the proportion of these costs is lower than those in vegetable cultivation. In the case of sweet potatoes, cocoyams and bananas all intercultural operations are done mechanically. Moreover, plantations of potatoes by mechanical means has also been initiated. Planting of sweet potatoes and cocoyams has been carried out partly by mechanical means. Considerable success has been achieved especially in harvesting some varieties of sweet potatoes with the help of potatoe harvesters. Bananas are harvested by hand and the transport of bananas inside the plantation area ist done mechanically.

Success in the Mechanization of Tobacco Cultivation

Production of tobacco is undoubtely an enterprise which needs the highest amount of labour per unit of cultivated area. In our country both dark and light coloured varieties of tobacco are produced. The first one covers 97.5 % of the total area under tobacco cultivation. Due to the requisite conditions for its production the mechanization of the light-coloured tobaccos has developed all over the world. We have utilized the experiences of the mechanization of the cultivation of the light-coloured tobaccos and with the help of broader research work developed models of machines for the mechanization of the dark-coloured tobaccos. At present one harvester which is capable of doing fullfledged harvesting is under trial. Under the present circumstances it is not possible to make a plan for complete mechanization because the fundamental problems arising from the shape of the soil surface and fragmentation and subdivision of land have still not been solved.

In general the cost per man-hour of manual work for harvesting accounts for 55-60 % of the total costs. In the context of future plans for reducing the labour costs, experiments are being carried out for establishing the complete mechanization of the harvesting of tobacco and also its transplantation, and for mechanizing irrigation and drainage systems.

Problems of Mechanizing Pasture Farming

In the pre-revolution period the pasture fields were scattered throughout the whole country. The vast areas covered by these pasture fields were not efficiently utilized. The abundance of cheap and surplus labour led to a low level of mechanization. During the present revolutionary stage many of the working processes were mechanized which brought about a considerable reduction in the labour costs. Now all efforts are concentrated on the mechanization of the planting of pasture fields and in some cases on mechanizing the transplantation in the pasture grounds.

Some Aspects about Qualification

We can mention here that some progress in the level of mechanization of the production of the above mentioned crops has been achieved. The progress which has been achieved in agriculture has been possible mainly as a result of the work done in the field of the educational system. These activities faced much difficulty at the beginning due to the high degree of illiteracy which we inherited from our pre-revolutionary regime. But the good work done by the Ministry of Education in the technological upbringing of our youth and adults and with that the simultaneous continuation of the process of qualification under the leadership of the National Directorate for Education and Propaganda of the INRA, have facilitated the achievements of the above mentioned success in mechanization.

In 1958 about 25 % of the population were illiterate who otherwise could have got education. The illiteracy could be practically removed be means of a campaign for the spreading of literacy among people. The constantly going process of professional education for the adult population was possible due to the farmer's and worker's schools and the farmer's and worker's faculties. During the last few years about 6000 to 7000 pupils and students were admitted every year into the professional and vocational schools of agriculture. The number of students in the agricultural faculties of the universities has also been steadily increasing. In the academic year 1973/74 3,123 direct students and 3,768 external students were registered in the universities and colleges.

In conclusion we would like to thank the organizers of this seminar for giving us the opportunity of speaking to you and hope that our humble contribution has provided some suggestions for you. We hope that we will achieve success in performing the task which is lying before us.

Table 1 The degree of mechanization of the different operations in the cultivation of the most important crops in Cuba

Crop	The present degree of mechanization %	Expected degree of mechanization in the future %	Reduction in the labour costs %
Sugarcane	32.0	80.0	48.0
Citrus fruits	42.0	63.0	77.4
Tobacco	22.0	39.0	25.3
Vegetables	15.2	69.8	30.8
Bananas	47.9	63.4	42.4
Pineapples	36.0	61.0	37.5
Pasture Fields	72.0	93.0	18.0
Cocoyam	31.0	36.0	8.8
Rice	97.0	97.0	6.4
Maize	62.0	90.0	71.7
Beans	61.0	83.0	65.5
Sweet Potato	46.0	63.0	37.7
Potato	70.6	89.0	87.5
Coffee (mech.)	15.0	29.0	17.4

Source: Informe de la Direccion Nacional de Mechanizacion a la Direccion del Sector Agropecuario, La Habana 1973.

Table 2 Progress in the Mechanization of Sucar Cane Harvesting

Crop Year	Proportion of Area harvested by combine harvesters to the total area %	Proportion of sugarcane area with mechanical loading to the total area %
1965	1.99	26
1966	2.97	45
1967	2.18	53
1968	3.44	61
1969	2.05	65
1970	1.00	85
1971	6.90	87
1972	7.00	89
1973	11.20	93
1974	20.00[x]	
1980	80.00[x]	

[x]According to plan figures.

Source: Revista ATAC. La Habana 1973, No. Julio-Agosto.

An Approach to Agricultural Engineering Development in Newly Reclaimed Land in the A.R. Egypt

by Ali Hossary

1. Introduction

One of the principal problems facing Egypt today is the very slow expansion of cultivated land area as compared to the very large increase in population. The Egyptian Government has attempted to meet this problem through various industrial and agricultural schemes as well as through a family planning programme.

Part of the agricultural programme is to reclaim about 1 to 3 million acres to be irrigated by the High Dam waters. To date about 800, 000 acres have been reclaimed at a cost of around Dollar 500 per acre. This cost includes engineering and grading the land, the installing the irrigation system, building roads, establishing administrative blocks and farming settlements, as well as carrying out initial leaching and establishing rehabilitation crops. Evergreen trees have been planted along the canals.

Once the main development has been achieved, the areas pass into the utilization and development stage. Most of the land is currently farmed as state enterprises.

The main organizational units are sectors, subdivided into farms of about 3000 to 5000 acres. It is, however, government policy to hand over much of the land to settlers. About 15, 000 families own blocks, and another 25, 000 have rented land. The holdings are 2-5 acres per family. The target of the state is to establish profitable balanced farming in the areas and then encourage community development to take over.

The Egyptian Authority for the Utilization of Reclaimed lands was established to develop new lands irrigated by waters of the High Dam. On the basis of the scientific technological cooperation between the German Democratic Republic and the Arab Republic of Egypt, a group of GDR specialists is assisting successfully in the development of agricultural engineering in the newly reclaimed land in the fields of seed production, cattle breeding, professional training, testing of farm machinery, and mechanization of agricultural processes. The Egyptian Authority for the Utilization of Reclaimed lands appreciates the sincere efforts of the GDR specialists to promote the development of agricultural engineering in all sectors. We feel that this fruitful cooperation has succeeded in strengthening the relations between both sides.

2. Stages of Land Reclamation

A soil survey was carried out on 14.5 million acres to select the soil to be reclaimed on the available water. It was found that only 8,328 acres are grade I land, 217,000 are grade II, 604,542 are grade III, and 1,391,682 are grade IV.+

Land reclamation passes through three stages:

2.1. Engineering stage

This stage includes levelling the land, implementing irrigation and drainage schemes, the construction of various types of internal canals and drains, construction works on the canals such as the building of bridges, wires, syphons, culverts, regulators and distributers, the locating of villages, public utilizations and road construction.

2.2. Agricultural stage

This stage concerned the treatment of the different soil types as well as increasing the fertility of the soil. It included two main steps:

a) Soil investigation

It is necessary to determine the physical and chemical structure of the soil in order to arrive at the most adequate type of soil treatment.

+ Classification according to soil fertility

b) Soil improvement

This stage aims at improving the mechanical, chemical, and biological structure of the soil. For instance, various methods of leaching or washing the soil are employed on salty land, gypsium is added in the case of alkaline soil, while the addition of mud or the growing of alfalfa is used to improve sandy soil. These steps are followed by cultivating the crops most suitable for the further improvement of the physical and biological conditions of the soil. As soon as the soil responds favourably to the crops, a suitable crop variation is applied in order to further raise the quality of the soil and to raise its productivity to the level necessary for subsequent improvement. Newly reclaimed land generally reaches an optimum level of production after three to six years of cultivation.

2.3. Social stage

Land which has been reclaimed, developed, and cultivated by the state, and which has reached a level of productivity sufficient for supporting a family on an average of five acres, is distributed to settlers from the neighbouring provinces who have met certain qualifications.

The farmers are settled in villages composed of about 200 houses. For every 300 families an agricultural cooperative society is set up to deal with seeds, fertilizers, equipment, and marketing.

3. Mechanization Development

The mechanized farms in the newly reclaimed land are good examples of areas where mechanization of agricultural operations has succeeded in reducing expenses by Dollar 16.25 per acre per year and has solved the problem of employing seasonal labour.

3.1. Agricultural equipment used

The agricultural equipment used in the reclaimed lands is extremely varied, but the main classifications are as follows:

a) crawler tractors: mainly used for tillage operations

b) main tillage equipment: 4-furrow mould board ploughs and rigid-tined cultivators (chisel ploughs)

c) row crop tractors, with provision for secondary tillage and planters, or seed drills

d) standard wheeled agricultural tractors, mainly of three makes in the 30-40 hp category (Some are used for field work but most are used for transport)

e) spraying units, some self-propelled

f) vineyard crawler tractors: 30-45 hp

g) harvesting equipment, including self-propelled combines, flax harvesters, swathers and binders

The reasons why a particular type of equipment is chosen are complicated. Generally, a deciding factor is the ease in obtaining spare parts, but the situation is complicated where land, labour, and capital may be the major limiting factors.

It would seem that the selection of crawler tractors for primary tillage in the newly reclaimed land unnecessarily complicates the mechanization. The land (except when dry and this can be avoided by irrigation) is not very difficult, and medium-powered wheeled tractors would have seemed likely to meet the requirements. Furthermore, although the acreages are large, the individual units of land are small because of the irrigation canals and drainage ditches. Thus large equipments are difficult to use advantageously, and in the case of tractlayers, there is frequent use of steering clutches and brackes. At this stage, it would be possible to change to medium-sized wheeled tractors in future planning.

Some of the problems that have arisen recently relate directly to the choice of equipment. In the long ran more attention needs to be paid to overall planning in the equipment selection

(a) towards universal usage and standardization

(b) efficient delivery of spare parts

(c) when bulk purchases are planned, work shop planning and staff training should also be simultaneously considered.

3.2. Mechanization facilities existing for reclaimed land

a) Service stations

They are established for each area of 1500 acres. They deal with preventive maintenance and the replacement of worn or broken parts.

b) General workshops

They are established for each area of 2500 acres and specialized in repairing tractor chassis and farm equipment. They comprise with a machine shop, a blacksmith, welding shop, carpentry shop, and stores.

c) Specialized workshops

They are fully equipped workshops and used to recondition complete standard exchange units such as engines, fuel pumps, gearboxes. Tractors and farm machines are nowadays constructed in such a way that it is quite easy to replace independant units or operating devices. Specialized workshops are servicing several sectors.

d) Training facilities

Training schools for drivers and mechanics bure have been established within the sectors. Each training school is capable of taking about sixty resident trainees. Demonstration rooms and small workshop facilities are available in each school. The main courses are one or two years for mechanics and one 14-month-course for tractor drivers. The mechanics, in their teens to early twenties, are selected among the settlers sons and are taught basic metal work and fitting techniques, and then proceed to tractor and vehicle repair. The tractor drivers are taught about a variety of machines including tillage implements, combines, and balers.

4. Agro-industries

With a view to establishing economically viable and socially integrated communities of settlers, the Government is now considering the introduction of agro-industrialization in its overall land reclamation and settlement programme.

The most promising areas for agro-industrial development, especially in the new lands, appear to be:

(a) expanded cultivation of horticultural products for export. Because of its climatic, geographical and labour situation, Egypt has comparative advantages over many other countries in the production and export of fresh fruits and vegetables.

(b) oil crops. The country is not self-supporting in fats and oils. Much of the newly reclaimed land is suitable for the large-scale cultivation of oil crops, the production of which could be integrated with industrial processing.

(c) industrial production of livestock products. The newly reclaimed land is cultivated with fodder crops for a number of years in order to improve its fertility. Much of the reclaimed land, owing to its physical qualities, appears to be suitable for the production of fodder crops, thus opening up good possibilities for large-scale livestock production.

The agro-industrial development is being initiated through a comprehensive programme of feasibility studies and measures for training the necessary staff.

The agro-industrial projects planned for the immediate future on newly reclaimed lands include:

(a) a tomato paste and tomato seed production plan, to be established in the Maryut reclamation zone. Its planned capacity is 2100 tons of tomato paste and 45 tons of tomato seed per annum.

(b) a grading and packing station for fruits and vegetables in the Tahrir reclamation zone to handle fresh fruit and vegetable for export and local consumption. The planned capacity is 15,000 tons of organges, 7000 tons of vegetables and table grapes and 300 tons of raisins.

(c) a meat-processing plant for the El Nahda settlement area in the north west delta reclamation zone. This plant is to be based on meat production to be undertaken in the El Nahda area. A nuclear herd of 1500 Holstein-Frisian cattle will be started. Units for artificial insemination, as well as plants for the mixing and processing of animal feeds and milk replacer composition for calves, will also be established.

The agro-industrial complex will be operated and managed by the settlers through "intermediate" cooperation societies, in cooperation with government agencies.

5. Agricultural Engineering Problems Related to Mechanization in Reclaimed Lands

5.1. Different physical surroundings

When machines are working under different physical conditions, for example a tractor designed for work in a cool country is transferred to a country like Egypt, to work under completely different climatic and other physical conditions, we have to recalculate all standards and norms, and the methods of testing and analysis. This refers to all units and aggregates, such as the cooling system, hydraulics, air and oil cleaners.

The standards applicable in the country of the manufacturer should be screened and investigated in order to find out to what extend they have been already adapted to the Egyptian conditions, and if not, what changes can be suggested.

5.2. Size of field and size of holdings

On the normal state farms, the traditional irrigation (basin or frequent field furrows) imported from the old land, renders mechanization of the harvesting extremely difficult. Thus irrigation requirements and methods have to be studied, together with a variety of problems associated with the size of fields and the size of holdings.

5.3. Fuel and lubricants

Fuels and lubricating material differ from one country to another, and the differences in quality are often remarkable. Especially the efficiency, which is closely related to the caloric performance of a fuel, can be different. If it is lower, the tractor will not reach peak performance. Another point to be considered is that impurities can lower the quality of a fuel or lubricant. Difficulties are to be expected when a tractor engine has been designed and built to run on a special type of fuel and the engine is expected to operate on other fuels instead.

5.4. Training of specialists

A fleet of machinery by itself does not ensure an increase in farm productivity. The decisive factor is the human being on whose knowledge and working experience the level of production depends. A planned agriculture based on science and advanced technology can progress only if there is a sufficient number of well-trained specialists. The mechanization of agriculture requires the active participation of agronomists, engineers, technicians, and machine operators.

The complicated machines and equipment utilized in mechanized agriculture require all personnel concerned to posses al-round knowledge. The training programme should include the fundamentals of engineering and agro-technology.

The creation of mechanized farming in the reclaimed areas makes it necessary to elaborate and disseminate among all staff working practically in agriculture, a complex of additional knowledge and experience to enable them to effectively utilize the mechanical implements and machines for the increase in production.

5.5. Organization of repair and maintenance

Attention and systematic care is required to keep the equipment in good working order, but these alone are not sufficient. Also needed is regular maintenance and skilled repairs. A general system of planned or scheduled maintenance must be introduced and used to keep the equipment in working condition at its rated productivity.

The system to be enforced should ensure

1. the maintenance of equipment for normal productivity and a reliable quality of performance
2. the prevention of unexpected or accidental breakdowns
3. an increase of productivity through improvements and modifications, made during overhauls, when applicable and desirable.

Agricultural Mechanization Development in Ghana. Past, Present and Future

by O. K. Gyarteng

1. Introduction

It is a well known fact that in most tropical countries, the mention of mechanization brings to mind, especially to the peasantry and laymen, the use of a tractor only. The recent appearance of trained agricultural engineers in the agricultural scene of tropical countries is probably changing this conception. Mechanization development should in fact apply to farming with simple tools such as the cutlass and the hoe, through the use of bullock-drawn implements right up to the most sophisticated tractors, harvesting and processing machinery. FAO (1971) defines mechanization as the study, manufacture, utilization, maintenance and repair of all tools, implements, machines and equipment which enable the farmer economically to raise the productivity of human labour.

In Ghana the cutlass (or the matchet) and the hoe have been the main agricultural tools for small-scale farming for a very long time. Originally, the local blacksmiths have produced most of them, until their importation (especially of the cutlass) was started by the large commercial trading firms which sprung up in British West Africa. They are now being manufactured in Ghana also. The cutlass has been much used in clearing the land for farming and planting, whilst the hoe has been best used for the preparation of the land, planting, and weeding. These tools are still predominant in Ghana's agriculture.

2. Animal-Drawn Implements

Records by Wadhwa (1964) show that the first attempts to develop mechanization in Ghana from hand tools to animal-drawn implements started in 1938 in the Northern part of Ghana, where most of the indigenous people reared domestic animals for a living. Ox-training centres were therefore set up at Tamale, Bawku, and Navrongo. Unfortunately, the interest in the new method subsided considerably for some time, probably due to lack of encouragement and due to the socio-economic conditions.

In 1951, however, the establishment of a Kusasi Development Cooperative Society brought new emphasis to the use of animal-drawn implements especially in the Kusasi District in Northern Ghana. Here farmers were granted loans to buy animal-drawn ploughs and in return the farmers were required to sell part or all of their groundnut harvest to the society.

There has been renewed interest in the use of these implements since 1966 when the Christian Service Committes (CSC) and the Government teamed up to assist the farmers. Ampratwum (1972) states that the Christian Service Committee, in addition to starting re-training of the farmers, developed a light ridging plough fitted with a yoke suitable for one ox (Kline et al. 1969). The National Institute of Agricultural Engineering of Great Britain also demonstrated in 1970 a multipurpose animal-drawn tool bar with a complete set of attachments in the Northern Region. The effect of the tool bar on the farmers, however, is yet to be assessed.

At the moment the demand for animal-drawn implements, especially the single furrow plough, is increasing. It is no wonder, therefore, that in addition to their importation from abroad, a local implement manufacturing firm is producing some of them for the market.

3. Tractors and other Farm Equipment

The tractor was introduced into Ghana's Agriculture in 1948-1949. This was at a large mechanized farm project at Damongo under the management of the Gonja Development Company, which was providing mechanical cultivation services to settler farmers in the district.

In 1950 tractor farms were set up by the Ministry of Agriculture at Pokoase, Kwadaso, and Nyankpala Agricultural Station, and in 1951 another farm was set up at Ejura Agricultural Station to investigate the problems of mechanization. In 1952-1953 mechanical cultivation services were introduced to farmers in various regions as part of farm machinery extension services by the Ministry of Agriculture. Here, tractors were hired out for field ploughing and carting of agricultural produce. The agricultural stations offered facilities for coffee hulling, oil-palm processing, maize shalling and rice threshing to the farmers. In addition, mobile rice threshers and hullers were sent out to farmers' compounds to process their crops. By 1954, grain driers were being introduced to the farmers (Wadhwa 1964, Ampratwum 1972).

In 1960-1961 the Agricultural Engineering Branch of the Ministry of Agriculture set up the Kpalaga mechanized cooperative farm near Tamale. At the same time a special seed-dressing extension service was introduced in the Northern and Upper Regions. For this project, 300 small and portable seed-dressing units were designed and fabricated in the Nyankpala Workshop by the extension staff to dress farmers' seeds.

By the beginning of the 1960s, the Government of Ghana had set up three production units, namely the Workers' Brigade, the State Farms Corporation, and the United Ghana Farmers' Cooperative Council (UGFCC). The Workers' Brigade was a paramilitary organization responsible for undertaking construction contracts and agricultural production. The State Farms Corporation, as its name implies, was set up to undertake large-scale farming for food and tree crops to feed the people and the industrial sector. And the UGFCC was instituted to implement the government's cooperative policy.

To this end, the Agricultural Wing of the Workers' Brigade from 1961 set up bash-clearing units which cleared, levelled, and prepared fields and established roads and drainage system at Somanya, Klikpo and Atebubu (Baffoe 1964). Agricultural stations of the Ministry of Agriculture together with their fixed assets were used as the nucleus for the formation of some state farms. At the same time, all the machinery of the Ministry of Agriculture was ordered to clear lands for the newly established state farms. By 1964 about 40,800 acres in 105 farms had been cleared for arable and tree crops.

In spite of all the attempts enumerated above, it was decided by the then government that the pace of development of a highly mechanized agriculture in Ghana was too slow. Therefore, the government embarked upon the importation of large numbers of farm machinery mostly from the USSR, Yugoslavia, and Czechoslovakia at the beginning of the 1960s. One major reason for the importation of the machinery was to make tractors available to farmers all over the country in an attempt to beat an 840,000 acre target set for the national Seven Year Development Plan of the 1960s.

The Government's initial order was for 1,000 tractors of Porsche, International Harvester, Deutz, David Brown and Massey Ferguson makes. These tractors started arriving in the country from early 1961, and they were shared among the state farms, the Workers Brigade, and the UGFCC.

In addition, the Workers' Brigade purchased some David Brown, Allis Chalmers, and TG-160 tractors and attachments between 1961-1963. The UGFCC also in 1963-1964 bought some trailers and harvesting equipment, whilst the Ministry of Agriculture purchased some caterpillars, balers, a combine, maize shellers and other equipment from the United Kingdom, and rice mills from the German Democratic Republic.

In 1961, the Government of Ghana signed a contract with the USSR for the supply of land-clearing tractors and other farm machinery to the state farms. The machinery supplied included C-100s, DT-54s mobile workshops and spare parts. Two other contracts were subsequently signed in 1963 and 1964 for setting up a mechanics' training school and the supply of rice and maize equipment for two state farms.

After importing some Zetor, BNT. 60 and TG. 90 Tractors in 1962, the UGFCC followed up in 1964 with the signature of contracts with M/S Motokov of Czechoslovakia for the supply of more Zetor wheel tractors and implements, and with M/S Rudnap and Agrovovojdina of Yugoslavia for supply for assembly of TG. 160 & 90 s and BNT. 60 crawler tractors with attachments in Ghana. In addition, for the resettlement of 80,000 people affected by the Volta River

Hydro-electric Scheme, the Government purchased 400 Massey Ferguson tractors,. with implements and some combines. The Sugar Products Corporation purchased 140 URSUS Tractors from Poland for the transportation of sugarcane from farms to the factory.

Kumar (1969) has listed (appendix I) the tractor population involved in Ghana's agriculture by 1961-1967.

Since 1966, there has been so much interest from various sides in farming that it has become extremely difficult to account for the number of tractors and the machinery sold to the private sector. This is because since 1966 most of the importation and sale of farm machinery has fallen into the hands of private firms or agencies. It is difficult to access how much has been achieved by the use of these tractors, however, a few recorded figures act as an indicator.

Wadhwa (1964) reports that the UGFCC mechanization services to the farmers started in 1963 and for that year 15, 000 acres had been prepared for farmers. In addition, the VRA Settlement scheme had prepared 1, 500 acres. By August 1964, 12, 760 acres clearing had been accomplished, 33, 323 acres land prepared for planting, and harvesting and transportation of farm produce undertaken for 40-50 thousand farms.

Nicholas (1964) also records that during 18 months in 1963-1964, 15, 000 acres of land were cleared for resettling 80, 000 people - flood victims of the Volta River Hydro-electric Scheme. Half of the land cleared was for rehousing and the other half for farming. On completion of the housing programme, all the machinery was engaged in land clearing for farming at the rate of 1, 500-2, 000 acres/month. To give each farmer a minimum of 6 acres, 54, 000 acres were cleared.

Appendix II lists the recorded acreages of clearing and cultivation accomplished with the UGFFC machines transferred to the Ministry of Agriculture for hiring to the farmers between 1967 and 1973. It is estimated that these figures are about 25 % lower than they should have been.

Irrespective of the shortcomings in the utilization of these machines, their involvement in Ghana's agriculture has made at least one mark - it has had an immense influence on both the farming community and the educated people who had hitherto not ventured into farming. The sad side of the story is that many firms and agencies have taken advantage of the increased interest of the people in agriculture by selling to them not less than ten different makes of tractors, many of which are grossly under-utilized due to a deplorable lack of aftersales service.

Other factors that attracted many people into farming were (1) the establishment in the 1960s by the Government of the Agricultural Development Bank which has since been granting loans to both small and large-scale farmers throughout Ghana, and (2) the announcement by the Government that public servants involved in agriculture would, on request, be granted permission to resign prematurely from the service to become full-time farmers; this resulted in the resignation of some public officers to establish tractor-hiring services and farms with loans from the Banks.

4. Farm Machinery hiring by the Ministry of Agriculture

The change of government in 1966 brought about the dissolution of UGFCC and the transfer of its assets (mentioned above) to the Ministry of Agriculture in 1967.

The Ministry has since then been rendering mechanization services mostly in land clearing, ploughing, harrowing, ridging, maize shelling, rotary mowing and combine harvesting of rice at highly subsidised rates. Appendix III shows the hiring rates as obtaining between 1967-1971 and new rates were introduced in 1971.

Again in 1971 the Government decided to give a chance to private enterprises in hiring wheel tractors to the farmers. To this end, the Ministry of Agriculture sold all its wheel tractors (with the exception of a few to be used as service prime movers) to the public. But the Ministry has maintained and hired out the heavy equipment for land clearing.

In view of the heavy expenditure involved in establishing a landclearing hiring service, it is obvious that it is going to take some time before privately owned land-clearing companies will develop on Ghana's agricultural scene on an effective scale.

In the meantime the Ministry of Agriculture is faced with the problem of meeting heavy demands for land clearing from farmers. This has become a major problem for the Ministry of Agriculture, because (1) the new Government policy of "Operation Feed Yourself" has not only inspired many farmers to increase their acreages but has also brought new hands into farming, especially rice farming in the northern parts of Ghana. (2) The Ministry of Agriculture's land-clearing service to the farmers is charged with the responsibility of working for the farmers only, and it has been the main source of land-clearing facilities since 1967.

(3) In spite of the high demand brought about by (1) and (2), the machinery strength of the Ministry is low and dwindling. It may be recalled that the machinery taken over from the UGFCC in 1967 had been imported into the country in 1962/63. One can imagine therefore the state of the machines now. The solution lies in the acquisition of new ones.

We are happy to note the great interest the Government is taking in improving the mechanization activities of the Ministry. Despite the heavy expenditure incurred by the Government since 1960 in procuring so much machinery, and despite the high expenditures year after year to try and keep these equipment running, and despite the present high cost of farm equipment and delayed deliveries, 51 rice combine harvesters (31 from the UK and 20 from Romania) were virtually rushed in from abroad to save from the threat of fire about 117,000 acres of rice farms ready to be harvested in the last three months of 1973. In addition to these combines, there are more than 40 combines owned by private farmers and corporations in the northern parts of Ghana alone, 80 more are expected within the year to forestall any combine shortage in the foreseeable future. Again there are ample assurances that the Government is acreeding to the demand for land clearing by importing more crawler tractors. The complete replacement of the land-clearing equipment is however to be phased out over a proposed five-year development period.

One may perhaps wonder at the need to import such expensive farm machinery when in a tropical country like Ghana there should be abundant supply of "cheap" labour to till the land, plant the seeds and harvest the proceeds. But this "cheap labour conception" no more applies to farm labour. Education, foreign influences, the apparently attractive life in the city compared with that of the peasantry, immigration controls and the hard life of tilling the soil under the scorching tropical sun, have caused the drift of farm labour from the rural to the urban centres. To the farmer, therefore, labour is becoming increasingly expensive and scare.

5. Irrigation

People from all walks of life inspired by the "Self Reliance" policy of the government of Ghana have been working voluntarily on several irrigation schemes in Ghana. The most popular schemes are "Dawhenya" to irrigate 1,200 acres, "Okyereko" to irrigate 300 acres and "Vea" to irrigate 3,500 acres. In all, the Ministry is presently undertaking 7 irrigation projects to irrigate a total acreage of 12,200 acres. It plans to start 16 others to irrigate 36,550 acres for which feasibility studies and, in some cases, detailed designs have been completed.

Separate from the above there are multi-purpose schemes to irrigate 238,000 acres of the Lower Volta Basin, about 8,000 acres near Weija, and the Avu-Keta Project to reclaim 74,700 acres in three stages.

6. Conclusion

Ghana has certainly had its mechanization problems in the past and many of them are still to be solved. But there have also been successes and some unique (though less known) examples of proper machinery, animal and labour utilization in several places.

However, with the present interest in improving mechanization and irrigation - with the allied needs of storage and marketing facilities - it is obvious that the emphasis in Ghana is on agriculture; and careful mechanization is a necessary fact of life. After all, tropical countries must have realised by now that the most effective way to develop is through increasing agricultural production and the provision of processing facilities for agricultural raw material in our own countries. We are hopeful of success, but equally the agricultural engineer in Ghana looks forward to the day the government takes a firm stand to standardize agricultural machinery imports into Ghana.

References

Ampratwum, D. B., A decade of agricultural mechanization in Ghana (1961-1970). Inter-Faculty Lecture presented on 16 th Nov. 1972 at the University of Ghana, Legon, Ghana.

Baffoe, W., Mechanized land reclamation pays the Workers' Brigade. Paper presented at the Seminar on Agricultural Mechanization in Ghana. Ghana Academy of Sciences and Ministry of Agric., Accra, Nov. 1964.

FAO: Mechanization of agriculture in tropical Africa. Paper presented at the Seminar on Mechanization in Africa, Bambey/Senegal, January 1971.

Kline, C. K., Green, D. A. G., Donahue, R. L., Stout, B. A., Agricultural mechanization in Equatorial Africa. Michigan State University 1969.

Kumar, H., Agricultural machinery inventory. Paper presented at the Symposium on Farm Mechanization, Ministry of Agriculture CSIR, University of Ghana, UST. Ghana 1969.

Nicholas, M. S. O., Heavy forest clearing experiences in the Volta River Authority. Paper presented at the Seminar of Agric. Mechanization in Ghana, Ghana Academy of Sciences and Ministry of Agric., Accra, Nov. 1964.

Wadhwa, N. D., Mechanical services to the farmers in Ghana. Paper presented at the Seminar on Agricultural Mechanization in Ghana. Ghana Academy of Sciences & Ministry of Agric., Accra, Nov. 1964.

Appendix I

Tractor Population in Ghana 1961-1967
(Kumar, 1969)

TEA 20	115
MF. 135	162
Fordson	2
MF. 165	306
Deutz	200
Porche	303
David Brown	224
Zetor	1400
MTS Belarus	34
Total	2746

Crawlers

C 100	109
DT 54	51
TG 160	12
TG 90	150
BNT 60	420
Caterpillars D 4	8
D 6	1
D 7	5
D 8	3
Euclid CD 8	2
Richard Continental	6
Allis Chalmers	1
Total	768

Let Westinghouse (Wheel)	5
Rice Combines	80
Rice Mills	26

Appendix II

Land-Clearing and Cultivation Statistics 1967-73
Ministry of Agriculture: Tractor hiring Service to the Farmers

Period	Acreage cleared	Acreage ploughed	Acreage harrowed	Acreage ridged	Acreage planted	Acreage slashing
1967 July-Dec.	900.45	8,821.03	5,978.30	613.80	90.72	175.00
1968 Jan.-Dec.	2,768.00	27,323.37	10,289.01	2,286.00	110.40	296.80
1969 Jan.-Dec.	-	-	-	-	-	-
1970 Jan.-Dec.	-	27,583.00	13,551.17	1,966.25	341.50	176.50
1971 Jan.-Dec.	-	4,718.85	1,433.50	613.25	50.00	304.00
1972^x Jan.-Dec.	1,034.06	5,112.78	1,674.45	55.50	49.00	73.50
1973^x Jan.-Sept.	1,106.00	4,913.11	1,576.00	264.50	4.00	229.00
Total	5,808.51	78,472.14	34,502.43	5,799.30	645.62	1,254.80

^x Figures on Ploughing and Harrowing not Available at Time of Printing

Acreage drilled	Acreage harvested	Acreage tree felling	Acreage windrowing	Acreage stumping	Acreage rooting/ ripping	Acreage ploughing/ harrowing	Acreage other operation (levelling earth work)
14.00	-	-	-	-	-	-	-
4.00	81.80	49.65	95.85	5.25	61.50	49.00	-
-	-	3,864.78	1,098.80	1,871.93	863.17	3,583.10	40.50
-	39.00	5,722.53	2,323.60	3,179.28	422.09	8,714.97	6.25
-	211.50	2,032.20	2,212.45	582.75	117.00	6,084.80	-
-	-	1,550.70	1,966.19	894.98	190.25	-	-
-	-	1,549.00	985.40	272.25	39.75	-	-
48.00	332.30	14,768.86	8,682.29	6,806.44	1,693.76	18,431.87	46.75

Appendix III

Subsidized Tractor Hiring Charges since 1967
Ministry of Agriculture

Description	Charges/acre 1967-71	Charges/acre since 1971
Clearing of virgin forest	Dollar 151.00	Dollar 56.50
Clearing of secondary forest	Dollar 97.00	Dollar 36.50
Clearing of costal thicket	Dollar 69.50	Dollar 29.00
Clearing of savannah woodland	Dollar 34.70	Dollar 13.90
Clearing of bush	Dollar 22.50	Dollar 10.40
Cultivation:		
Ploughing	Dollar 3.50	Dollar 5.20
Harrowing	Dollar 1.75	Dollar 3.50
Ridging	Dollar 3.50	Dollar 3.50
Inter-Cultivation	Dollar 1.75	Dollar 1.75
Planting or Drilling (with or without fertilizer)	Dollar 4.35	Dollar 4.35
Slashing	Dollar 3.50	Dollar 3.50

The Development of Agronomic Practices for the Production of Field Crops under Irrigated Conditions

by Ali A. El Khishen

1. Introduction

Man learnt to select and cultivate his food plants thousands of years ago. Ever since, he has been trying to develop methods and techniques to cultivate larger areas and to obtain better yields. His objectives in doing so were as follows:

- to save some seeds from the selected plants and keep them until the time of sowing
- to turn over the soil and prepare a suitable bed for sowing the seeds
- to clean off the unwanted plants (weeds) that grow and compete with the crop plants
- to protect the growing plants against pests
- to harvest the crop and store it to be used throughout the year
- to use tools and equipment that hasten farm operations and save time, effort and cost of production.

These objectives are exactly the same as those of the cultural methods used nowadays, and they will remain the same for the farmer of the future.

However, the means and methods of execution are always improving; first by the efforts and the ingenuity of the farmer himself and later by organized and well designed agricultural research.

The improvement of cultural methods over the centuries has gone through three main steps:

(1) manual farming, when man used physical power to cultivate his crops

(2) Man-and animal-power farming, when the farmer was able to domesticate animals, like horses, mules, oxen and buffaloes as powerful means to

help him cultivate larger areas and to do the most difficult operations especially ploughing and transportation. In doing so, man invented new and more developed tools to suit the new power.

(3) mechanized farming in which the man-power was mostly saved and animal-power was completely replaced by tractors.

2. The present situation

If we take a look around the world of agriculture we find that the advanced and more developed countries have gone very far in the use of modern farm machinery and technology. By adopting this system of mechanized farming these countries have not only cut down the cost of production and increased the yield of field crops, but also released much of the human effort for the advancement of other activities. It may be stated that the use of farm machinery and mechanized farming are the backbone of modern labour saving life, in which agriculture has become an occupation and industry, as respected and advanced as any other industry.

On the other hand the production of crops in the less developed countries still depends to a very large extent on both man- and animal-power, without any improvement taking place. The Egyptian farmer, for example, has been ploughing his land and producing his crops, by almost the same methods and techniques as those used by his ancestors five thousands years ago. Any progress he has achieved lately is only very slight and limited. Farm tractors however, have found their way to a limited extent, to the developing countries, but their use and efficiency is rather low. It is questionable if the way of handling and operating these tractors is correct and scientific. As for other farm equipment, one cannot see any serious efforts to make any large-scale use of them. They comprise sowing machines like seed drills and planters, cultivation and fertilizer equipment, tractor-mounted field sprayers and dusters, harvesters like combines, corn pickers and forage cutters and mowing machines with irrigation, the Egyptian farmer has developed his own methods adapted to crop and animal farming. His ploughing is rather shallow and does not turn the soil and trash under. The plough does not go deeper than 10-12 cm, thus ploughing has to be repeated twice or three times in order to obtain a good seedbed. He then divides his field into either ridges or small plots of about 5 x 10 m to facilitate control and irrigation. The farmer usually keeps a slope of 15 - 25 cm per hundred to make sure that water can reach all parts of his field by gravity. This method is rather costly since it consumes much more water than is actually needed for growing crops. In sowing, the seed is either broadcast sown by hand or manually placed, the position of the seed being far from suitable for good germination. Thus the farmer uses two or three times more seed than by mechanical seed drills or planters. Weeding, the most laborious among the farming operations, is done entirely by hand. Since it is very important to keep the field clean, the cleaning must be repeated several times, as new amounts of weed seeds are added with each irrigation. The irrigation water becomes heavily contaminated with seeds of many weeds growing thickly on the banks of the irrigation ditches. Pest control is also done mostly by hand, the use of knapsack sprayers or-long-hose motor spayers do not guarantee the efficient or uniform distribution of the pesticides throughout the field. Herbicides are seldomly used, since some of them have to be quickly incorporated into the soil in order to be effective. All the pesticides and herbicides must be applied in a very accurated and precise manner and at the recommended doses in order to destroy the pests without damage to the crop plants. This cannot be done by hand; tractor-mounted sprayers are specially designed for the job. The harvest operation in the developing countries is rather complicated, it is usually done by hand, consuming much time and human effort. In most cases labour is not sufficiently available at the proper time, so that the operation goes on very slow, resulting in big losses through scattering and the birds and mice. Handling the crop during harvesting and threshing incurs further losses. For example, in Egypt in 1958 up to 22 per cent losses occured when wheat was harvested by hand and threshed by animals, while only about five per cent losses occured when self-propelled combines were used. About fourteen per cent lossed occured when the harvesting was animal powered and threshening done by a stationary thresher.

3. Future possibilities

To help the developing countries improve their agricultural production, research should play a major part. New agricultural technology should be adopted by these countries within scientific and well-planned programmes. Research has achieved many outstanding results, especially in plant raising with countries enjoying the use of high-yielding varieties of wheat, hybrid maize, rice, sorghum, and many other crops. Other research findings were obtained concerning crop production like the times and rates of sowing, population density, weed and pest control, fertilization and plant nutrition, irrigation, application, harvesting and handling techniques, but most of these findings can hardly be used under the traditional farming methods. Consequently in the developing countries research work should include the introduction of modern farm machinery as a major objective to support the farmer and help him to produce more and to give him a higher living standard.

With this idea in mind the Agronomy Department of the Faculty of Agriculture, University of Alexandria, has begun together with the cooperation of the Ministry of Agriculture, a series of studies and trials to mechanize the farming operations used in the production of major crops, namely cotton and maize as row crops and wheat, rice and berseem clover as broadcast or drill crops at field station, such as Mehalet Mousa and Skha and Kafr El-Sheikh.

These studies started in 1964 in which a 6-acre field in the University of Alexandria Farm and a 4-acre field in Mehalet Mousa Farm were planted mechanically for the first time with cotton. All farm operations were executed by the newly introduced equipment - except for the picking, which was done by hand because no precise cotton-picking machine for the fine, long-staple Egyptian cotton has been invented as yet. The results were encouraging and allowed for more elaborate studies to compare the traditional methods of cotton production with the newly introduced mechanized techniques. In one of these comparisons, 200 acres of cotton, prepared, sown and served by hand and animal labour, were compared with an adjacent 200-acre plot of the same variety of cotton planted and served by machines. The results showed about 25 per cent cuts in the cost of production and about 35 per cent increases in yield.

In order to use this modern machinery under irrigated field conditions in Egypt, the ordinary small field plots should be changed into long stretches of land free from interfering irrigation ditches, levees, or drainage canals. Thus different lenghts and widths of the irrigation plots were investigated. Levelling was also important, so different degrees of slopes were investigated. Direct flood versus syphon tube application of irrigation water were compared, both in flat and ridged fields. These investigations have shown that the length of the field may vary between 200 m and 300 m, with a slope not less than 10 cm/100 m and not more than 400 cm/100 m. This length would allow the tractor runs to be more economical with the least number of turns. If the length was more than 300 m, the flow of irrigation water would be less uniform resulting in uneven distribution of water throughout the field plot. The width of the plot was found to depend upon the kind of crop since some seeds like cotton need more water for germination than maize, wheat or berseem clover. Thus the most favourable range of plot width was found to be from 10 - 12 m for wheat, flax and clover to about 30 - 40 m for cotton.

The investigation of the mechanization of the major crop production in Egypt in the last ten years has given quite satisfactory results which may be summarized as follows:

1. A considerable decrease in the cost of production has been achieved.
2. A sizeable increase in yield resultet

This may be due to:

(a) the high efficiency of the modern equipment in performing the farming operations exactly according to the scientific recommendations concerning crop management
(b) better timing for sowing, cultivation, pest control, and harvesting. This is very important especially in a fast crop rotation, like in Egypt.
(c) the methods of mechanical harvesting, which are designed to prevent or minimize the losses that are usually high in the manual handling of crops.

4. Recommendations

After this survey of the situation the following recommendations may be made for the improvement of food crop production:

1. The mechanized use of crop production must be a major goal in the developing countries.
2. The agricultural scientists should include in their research programmes the necessary studies for the adaptation of the proper farm machinery to suit local conditions.
3. Agronomists and agricultural machine engineers should cooperate in these studies.
4. Government institutions should be called upon to establish farm machinery stations throughout the country to execute the major field operations mechanically for the farmer at a reasonable price. These stations should be run by competent agronomists, agricultural engineering technicians, skilled labourers etc.
5. Extension work should play an important role in the mechanization programme. Some of the extension workers should be specialised through training courses to get all the information about the applicable results of the research work.

Notes on the Mechanization of Dairy Farms in the A.R. Egypt

by M. T. Ragab

Mechanization in the field of animal production in Egypt (animal husbandry) has only very recently been thought of, because cows and buffaloes are usually owned in small numbers (1 - 2 head per owner). This situation, from another point of view, has indirectly promoted mechanization. The scattered ownership was and still is a handicap in collecting milk from the farmers in a feasible trading size, and dairy factories have been complaining of a shortage of milk. Therefore farmers were persuaded to build large herds of dairy cattle, or to take up intensive dairy farming. Also, with the development of the newly reclaimed land, a great number of large dairy farms was set up in order to complement animal and plant crops, and to solve the problem of milk deficiency. The size of any of those dairy herds was 300 heads of milking animals, then increased to 500 heads, over and above the followers. In the course of implementation certain problems were faced: (1) the inadequacy of skilled labour to attend and milk those animals, (2) the high capital expenditure per milking head, (3) the requirement for enough cold storage to keep the milk for at least 24 hours or longer, since daily collection is not guaranteed.

Therefore, mechanical milking was adopted along with the supply of cold storage to those farms able to store two days' milk production (three tons). Different systems of machine milking were used, milk line, tandem and herring-bone systems. Data were collected on the behaviour of the animals and the efficiency and economy of the systems. The results have been very encouraging, but steps had to be taken in order to reduce the costs. The situation is a result of the great numbers of followers, which are non-producing by nature and of the relative proportion of the milking heads to the total number of the herd. Therefore, the buildings are used for the largest possible numbers of milking animals, and the newly born calves only to the age of weaning, when they are transported to other locations where less expensive housing and cheap feeding is available. The solution was not easy, since it increased the milking herd three times from 500 heads to 1500 heads on a limited area of land. The next step was the idea of mechanizing other operations, such as feeding and manure disposal. This is being implemented in one of the farms and, depending on the results of this experiment, the modification of other farms may follow. In our organization we have 17 dairy farms, 15 of which are mechanized in milking along with two others in the newly reclaimed areas.

One of the problems which faced was milking the buffalo by machine. In our country the buffalo is the main native dairy animal, and it is a good source, not only of milk, but also of a high fat percentage (average 6-7 %).

Improvement in this animal was mainly concerned with the milk yield; no efforts were made for the type. When we adopted the idea of machine milking we were confronted with some handicaps: the animals were not accustomed to milk without their young beside them.

They must be massaged by hand for a relatively long period before giving their milk. They frequently hold the milk back and have to be massaged again before they let down the milk. The teats are of many different shapes and not suitable to the machine, since they are not even in length, thickness, shape position and development. However, the G.D.R. milk processing industry (Impulsa) did their best to minimize this difficulty, and we selected the animals with the least inconvenient characteristics.

It is worth saying that the buffaloes at the beginning were rather non-responsive but with patience and training of both the milker and the animals outstanding results were achieved. As far as mastitis is concerned, it may be stated that machine milking did not cause more incidences than in the case of hand milking. The time required was a bit more at the beginning than needed for cows per unit of weigth and time, but now the difference is no longer significant.

Finally, it should be emphasized that mechanization is the only solution to the various problems of intensive production, which is the pattern of our age. Therefore, it ought to be given its due share from both the engineer and the breeder. It will not be a long time before partially or fully mechanized dairy farms will be the general system, since they ensure the control of many environmental aspects, along with the more rigid control of management, over and above the protection of animals from so many diseases through more or less complete isolation from contamination.

Actually, we are now in a new era of animal husbandry, where the machine will be a fundamental part of the farm.

The Need of Training Manpower for Mechanized Agriculture in Iraq

by M. Rafi, H. F. Jumah, L. K. Ismail and M. Asrar

1. Introduction

Lately in developing countries, the number of tractors and modern farm machinery has been on the increase. There is also a growing awareness among the developing nations of the role which agricultural engineering technology, especially agricultural mechanization, can play in increasing agricultural productivity and improving rural life. However, the facilities for the education and training of personnel at all levels to man this new technology in agriculture are lagging far behind this trend. Thus in many instances the planning, implementation, and operation of agricultural mechanization projects, both in public and private sectors, is entrusted to the hands of personnel who have either no or inadequate formal education and training in the field of agricultural mechanization. The result is that instead of reaping benefits from the mechanization of agriculture, these countries are bearing colossal financial losses on this type of investment. This gap between the need and the available manpower for mechanized agriculture is becoming a serious problem, and considerable efforts are necessary to remedy this state of affairs.

Before dealing with this problem in Iraq, some of the relevant concepts should be outlined. Hitherto, agricultural mechanization has been taken to imply only the use of modern machinery in agriculture including horticulture and animal husbandry. In fact, agricultural mechanization cuts across all activities relating to agriculture and rural life and extends the application of machines and/or relevant engineering methods and practices to: (1) production operations in agriculture; (2) irrigation and drainage of agricultural lands as well as soil and water conservation; (3) processing, handling, and storage of agricultural products; (4) farm service buildings, rural housing and roads; (5) electrification in agriculture, and rural areas. It should be noted that agricultural engineering deals with design and development, i.e. analytical aspects of the above-mentioned activities, whereas agricultural mechanization covers the application and utilization aspects of machines, methods and practices which have been developed by agricultural engineering.

2. The present situation of Iraqi agriculture

After crude oil, the agricultural sector is the second important source of Iraq's economy. Since 1968, the Revolutionary Government of Iraq has given priority to the development of this sector in order to achieve self-sufficiency in food and fibre and at the same time raises the standard of living of the masses. In view of the probable world food shortage in the years to come, this sector has to be mobilized in all its dimensions to meet this challenge and to relieve the country from dependence on foreign sources for certain food and fibre commodities.

The total area of Iraq is 43, 844, 600 hectares. It is estimated that 27.4 % of this area (12, 000, 000 hectares) is culturable agricultural land, the remaining land being deserts, mountains, lakes, marshes, etc. The culturable agricultural land includes 8, 000, 000 hectares of irrigable land and 4, 000, 000 hectares of land in the rainfed region. At present, only 50 % of the irrigable and rainfed lands are cropped annually. It is estimated that by further haernessing water of the Twin rivers (Euphrates and Tigris) nearly 75 % of the irrigable land could be brought under farming.

The southern region of Iraq has an arid climate where only irrigated agriculture is possible on culturable land. The climate of the northern region is semi-arid where agriculture depends on natural precipitation. Some irrigated agriculture is also practiced along the banks of the river Tigris and its tributaries in this region.

The main problem facing agriculture in the southern region is that of salinity and water logging. On the other hand in the northern part, water and soil conservation practices need immediate enforcement for enhancing agricultural productivity and maintaining soil fertility.

The four principal field crops of the country are wheat, barley, rice, and cotton. Dates and grapes are leading fruit crops. The areas under field crops, vegetables, and fruit trees constitute approximately 90 %, 5 %, and 5 % respectively of the total cropped land. Among the field crops, the grain crops occupy about 85 % of the total cropped area.

An interesting phenomenon which is taking place in rural Iraq and which is going to influence and accelerate the pace of mechanization of the agriculture, is the drift of agricultural labour to urban areas (table 1).

Table 1 Population statistics of Iraq

Year of census/estimates	Urban	Rural	Total
1965 (census)	4, 162, 106	3, 935, 124	8, 097, 230
1972 (estimates)	6, 065, 000	4, 009, 000	10, 074, 000

It can be predicted that this migration will ascend in the coming years in view of the revolutionary policies of the present regime of Iraq which aims at providing basic industries within the country.

From this discussion of the situation of Iraqi agriculture, it may be said that it presents a classical setting for mechanization of the agriculture. This situation has arisen from: (1) the drift of agricultural labour to urban areas; (2) the ease with which 85 % of the total cultivated land under grain crops can be mechanized; (3) the need of applying engineering principles and practices in solving soil and water problems; (4) the national goals of achieving selfsufficiency in food and fibre by utilizing all modern means of agricultural production and rural life.

3. Present status and future scope of agricultural mechanization in Iraq

Table 2 indicates the rate at which agricultural mechanization has developed in Iraq during the last 15 years. (The figures have been estimated from different valid sources).

Table 2 Agricultural mechanization situation in Iraq

Item of agricultural mechanization	Numbers in years		
	1958	1970	1973
Tractors	2, 400	10, 400	12, 464
Farm implements and machines	2, 600	12, 745	15, 000
Combines	1, 000	2, 280	2, 669
Irrigation pumps	5, 400	9, 150	16, 500

The farm machinery is owned both by the state and private sectors. The utilization of farm machinery in the state sector is administered through the Farm Machinery Hiring and Repairing Stations.

The total number of such stations in Iraq is 19. The sales and the services of farm machinery for the private sector are also managed by the state. These services are being further organized, developed, and expanded with the objective of extending their availability throughout Iraq.

The authors hold that Iraq has presently the equivalent of 0. 24 hp (human + animal + mechanical) per hectare of cropped area[x] while the minimum need for achieving a reasonable level of productivity in agriculture is considered to be 0. 5 hp per hectare. This target can be best reached in Iraq by increasing the number of tractors, combines, and irrigation pumps. To meet this requirement, bearing in mind the drift of agricultural labour to urban areas in future, the number of tractors, combines, and irrigation pumps by the end of this decade should be three times the present level. The above estimates are also supported by the targets worked out by the ministries concerned which are aiming at having 32, 000 tractors, and 8, 000 combines by the end of the current decade. In order to match the agriculturally advanced countries in the world Iraq would have to go beyond the minimum need of power per hectare in the coming decades.

There are three basic components of continuous agricultural mechanization, the energy, national farm machinery industry, and the trained

[x] The calculation for this estimate is given in the appendix to this paper.

manpower at all levels for this programme. Nature has bestowed Iraq with abundant energy resources. The present regime has recently established at Iskandariya the National Farm Machinery Industry, catering for the needs of mechanized agriculture. This industry, though in its infancy, is progressing in the right direction. The third and the most important component, the trained manpower at all levels for mechanized agriculture, needs immediate attention by the authorities, as the success or failure of the whole programme depends on it.

4. Survey of existing educational and training facilities

A review of the existing educational and training facilities in agricultural engineering and agricultural mechanization in Iraq with suggested guidelines for remedying the shortcomings, is valid at this point of discussion.

4.1. Higher level

Colleges and universities can offer higher level courses in agricultural engineering and agricultural mechanization. As undergraduate courses have to be established and developed before any post-graduate programme is offered, we shall concentrate on the undergraduate programme. Undergraduate education in Agricultural Engineering and Agricultural Mechanization at college or university level in Iraq comprises:

(i) B.Sc. in Agricultural Engineering
(ii) B.Sc. in Agricultural Mechanization
(iii) Service Courses in Agricultural Mechanization for students of agriculture and related disciplines

(1) B.Sc. in Agricultural Engineering

The Colleges of Engineering at the Universities of Baghdad and Mosul are presently offering first degree programmes in Agricultural Engineering at the professional engineering level. The first graduated in 1971 from Mosul University; the University of Baghdad will turn out the first group of graduates in 1974.

The Departments of Agricultural Engineering at these universities are in their initial stages of development and require teaching staff and well-equipped laboratories, which are so essential for producing first quality agricultural engineers. At present, these departments are partly staffed by foreign teachers, which is no permanent solution.

Iraqi agriculture and its related industries offer a host of opportunities of employment for qualified agricultural engineers. These graduate engineers would be particularly needed for design, development, testing, and analytical work in the profession.

(2) B. Sc. in Agricultural Mechanization

There is an urgent need of agricultural mechanization technologists to man agricultural mechanization projects at the planning, implementation, and extension levels. These graduates are required to work as extension, advisory, and administration personnel in governmental and semi-governmental organizations dealing with agricultural mechanization projects, sales and service personnel in commercial farm machinery firms; progressive mechanized farms. Qualified in agriculture and agricultural mechanization, such graduates will also be proficient in agricultural production technology. As such they would prove highly effective and useful agricultural extension workers.

However, no academic institution in Iraq is at present offering such courses. They should be started as soon as possible in the two well-established Colleges of Agriculture in Iraq (at Abu Ghraib and Hammam Al-Alil).

(3) Service Courses in Agricultural Mechanization

Every graduate of agriculture and related disciplines must have adequate instruction in agricultural mechanization relevant to his field, so that he is conversant with the available machines and methods which he will be dealing with during his professional practice. Such courses should form an integral part of all the curricula in agriculture and related disciplines.

Except at the College of Agriculture and Forestry at Hammam Al-Alil, the number and the contents of these courses at the other Colleges of Agriculture would require expansion and development. The Colleges of Agriculture in Iraq have the same problem of inadequacy of qualified Iraqi teaching staff and well-equipped laboratories as the Departments of Agricultural Engineering at the Colleges of Engineering. This situation demands the active attention of the concerned authorities on a priority basis.

4.2. Intermediate Level

Intermediate-level technicians in Agricultural Mechanization come in between the agricultural engineers agricultural mechanization technologists, and the skilled workers, and their importance for the success of the whole mechanization programme should not be underestimated. Hundreds of such technicians would be required for the introduction and implementation of mechanized agriculture in the country.

At present, the Institute of Agricultural Technology at Abu Ghraib is the only institute which offers intermediate-level programme in agricultural mechanization. The existing curriculum of this institute would also need some revision in order to increase the scope of the training. It would also be desirable to name this programme "Agricultural Mechanization Technician Course".

The Government of Iraq is planning to establish four more such institutes at Kufah, Maysan, Baquba, and Eske-Kalak in the near future. These four additional institutes would be able to train the intermediate-level workers required for mechanized agriculture in the country.

4.3. Vocational Level

Vocational training in agricultural mechanization would cover the training of farm machinery operators and mechanics. Presently, no formal training programme is available for this group of skilled workers. In view of thousands of tractors, combines, and other farm machinery which are in operation in rapidly increasing numbers, the institution of such training courses at the appropriate centres is a must.

It is suggested that these training courses can easily be institued at the existing 19 farm machinery hiring and repairing stations, with the provision of requisite additional facilities. These stations should offer both regular courses for new trainees, and short in-service training courses for qualified farm machinery operators, and mechanics. The Ministries of Industry, Agriculture, and Agrarian Reforms can share the financial responsibilities of these training courses.

5. Trained Manpower Requirements

Based on the estimates of the ministries concerned, which are on the conservative side, as to the number of tractors and combines in Iraq by the end of 1971-80 decade, the numbers of trained personnel at various levels required for this purpose are given in table 3.

Table 3 Trained manpower requirements for agricultural mechanization in Iraq by the end of 1980

Category of manpower	Numbers needed by the end of 1980
A. For farm machinery activities	
1. Agricultural Engineers/ Agricultural Mechanization Technologists	400
2. Agricultural Mechanization Technicians	800
3. Farm Machinery Mechanics	4,000
4. Farm Machinery Operators	40,000
B. For other agricultural mechanization activities	
1. Agricultural Engineers/ Agricultural Mechanization Technologists	1,200
2. Agricultural Mechanization Technicians	2,400

There will be one operator, one mechanic, one supervisor/foreman, one agricultural mechanization technologist/agricultural engineer for 1, 10, 50, and 100 self-propelled agricultural machines respectively. This ratio is based on the authors' experience.

It should be noted that these requirements are only one aspect of agricultural mechanization, namely farm machinery activities. More personnel would be required for providing other aspects of agricultural mechanization. In modern farming, the agricultural mechanization is not only the link between the pure agricultural sciences and their application to crop production operations on the farms, but it also cuts across all activities concerning agriculture and rural life.

In the opinion of the authors, at least one agricultural mechanization technologist/agricultural engineer, and two agricultural mechanization technicians would be required for other aspects of agricultural mechanization for each agricultural cooperative in the country (table 3). There were 992 agricultural cooperatives in Iraq in 1972. Each agricultural cooperative covers an area of 2,500 hectares on an average.

It is strongly held that the above are the minimum requirements of such manpower. They will increase manifold in the coming decades. There are two ways for organizing the training of the required manpower. The first way is by instituting regular training programmes and extending the period of achieving the targets beyond the year 1980. The second way involves more intensive training programmes with the objective of achieving the above targets by 1980. This suggestion comprises two-stage training activities, i.e. in addition to the regular training programmes as given under choice I above, special short-term condensed conversion study programmes of about one year should be offered. These short-term training programmes should be designed and developed for graduates in agriculture and engineering, and diploma holders from technical institutes.

6. Conclusion

It is evident from the above discussion that there is an acute shortage of trained manpower at all levels for mechanized agriculture in Iraq, which is so essential for successful and continuous mechanization in the country. Enormous efforts are needed to remedy this state of affairs. In this connection, the following guidelines are suggested:

1. An action programme of higher education and training abroad for Iraqi personnel should be worked out and got underway, so that Iraqi personnel is then available in the country to take the lead in education, research, extension, testing and development, and planning work related to agricultural engineering and agricultural mechanization.
2. The Departments of Agricultural Engineering at Baghdad, and Mosul, should be further strengthened both in respect of staff, equipment, and other facilities.
3. B.Sc. courses in agricultural mechanization should be established at the Colleges of Agriculture at Abu Ghraib, and Hammam Al-Alil at an early date.
4. The number of service courses in agricultural mechanization should be increased in the curricula of all Colleges of Agriculture in the country. A good initiative has already been taken to this effect in the College of Agriculture and Forestry at Hammam Al-Alil.
5. The agricultural mechanization technician course offered at the Institute of Agricultural Technology, Abu Ghraib, should be revised in order to increase the scope of the training.
6. The process of establishing four additional Institutes of Agricultural Technology similar to the one at Abu Ghraib should be accelerated so that the much needed intermediate-level workers in agricultural mechanization can be trained.
7. Training courses for farm machinery operators and mechanics should be instituted at all farm machinery hiring and repairing stations with the provision of requisite additional facilities. The expenditure on the conduct of these courses should jointly be borne by the Ministries of Industry, Agriculture, and Agrarian Reforms.

References

Al-Rawi, H., Al-Khaffaf, A.M., Views on agricultural mechanization. A short study in the development of agricultural mechanization in the Republic of Iraq between 1966-1970 (Arabic). Al-Gadha Publishers, Najaf, Iraq 1971.

Buringh, P., Soils and soil conditions in Iraq. Ministry of Agriculture, Republic of Iraq, Baghdad 1960.

Giles, G.W., Towards a more powerful agriculture. West Pakistan Department of Agriculture, Lahore 1968.

Republic of Iraq, Annual abstract of statistics. Central Statistical Organization, Ministry of Planning, Baghdad 1972.

Rafi, M., A guide for planning and developing agricultural engineering and agricultural mechanization education programmes and supporting facilities at agricultural colleges and universities in developing countries. College of Agriculture and Forestry, Mosul University, Hammam Al-Alil, Iraq 1973.

Personal visits of the authors to the concerned Ministries, Colleges, and Institutions in Iraq.

Appendix

Estimate of the horsepower per hectare available for crop production in Iraq at the end of 1973

1. Agricultural area		
Cropped area including permanent crops like fruit trees and fallow cultivated area	...	8, 000, 000 hectares
2. Human labour		
Taking agricultural labour equal to 1/5 of the total population of 10, 396, 000 (1973 estimates)	...	2, 079, 200 numbers
1/15 hp each	...	138, 613 hp
3. Work animals		
Taking estimated heads of 1, 100, 000 work animals comprising horses, mules, donkeys, and camels	...	1, 100, 000 numbers
0. 5 hp each	...	550, 000 hp
4. Mechanical power		
i) 12, 464 tractors 50 hp each	...	623, 200 hp
ii) 2, 669 combines 75 hp each	...	193, 575 hp
iii) 16, 500 irrigation pumps 25 hp each	...	412, 500 hp
		1, 917, 888 hp

Therefore, horsepower per unit of cropped area $= \frac{1,917,888}{8,000,000}$

= 0, 24 hp per hectare

Towards a Plan for Agricultural Mechanization in the A.R. Egypt

by Mohamed Mahmoud Abdel Raouf

I. Introduction

A well known truth about the agricultural development that was accomplished in all developed countries is, that this development has been brought through taking advantage of the achievements of modern technology and science in the domain of agricultural production. One of these achievements is the production of specialized machines for use in the agricultural production.

In order to introduce mechanization in its scientific form and to make a plan for mechanization of Egyptian agriculture, we have to take into consideration the objective conditions of Egyptian agriculture. Some of these conditions are farm size, availability of necessary cadres, availability of motive power. Special consideration should also be given to the problems of manpower in agriculture and their relationship to the problems of manpower on the national level. Attention must also be given to the relationship between mechanization and the efficient utilization of the more rare elements in the agricultural sector like cultivated land, livestock and water. Furthermore we must give more attention to the costs and benefits of mechanization in money terms and social costs. The way and time limitations of the execution of such a plan must be taken into consideration. All these problems must be regarded in order to answer the following questions:

a) Is the mechanization necessary for Egyptian agriculture?
b) How to put a plan for mechanization?
c) Who wants mechanization?
d) Who needs mechanization?
e) What is the direct and indirect effect of mechanization on the economic and social conditions in rural areas and on a national level?

II. Mechanization and the objective conditions of Egyptian agriculture

1. Mechanization and the efficient use of cultivated land in Egypt

It is important to look upon cultivated areas as a national wealth. The cultivated area in Egypt is very limited and its horizontal expansion is also limited with a high expenditure. Then we must use it more efficient to attain the maximum production and at the same time increase its fertility. Field experiments and practices show that the cultivated area in Egypt will produce more at a lower cost.

2. Mechanization and the efficient use of animal wealth

The field experiments and practices show that the replacement of animal traction-power by machines will save a great amount of milk and meat, i.e. about 700 million kg milk and about the double of the to-days' meat production in Egypt. When we attain only half of the quantity of milk and meat, it is reason enough to substitute animal work.

Also the 1, 5 million feddan under fodder will be utilized to feed high milk and meat yielding animals instead of the low yielding buffeloes and cows.

3. Mechanization and efficient use of water resources

The Egyptian agriculture is consuming most of the irrigation water from man-made irrigative system. This is due to many factors. The farm size plays an important part. The Egyptian farmers consider water to be a commodity without price and therefore they increase their condumption of water, using primitiv irrigation techniques, like great length of irrigation sub-ditches and canals and long distances are breached by water between the main irrigation canals and the land to be irrigated.

To get 1 m^3 water we need L.E. 0.002 as fixed costs in the High Dam. This costs will increase to L.E. 0.0036 by the execution of Middle-African projects in the basin of the river Nile. To get 1 m^3 we must pay also L.E. 0.002 as current costs (i.e. cost for the ministry, organization supervising the river Nile and being responsible to supply water into the canals prepared by the farmers). Furthermore water is the strategic element in the future of Egyptian agricultural horizontal expansion. The more we economize the use, the more we save it for agricultural expansion purposes. By mechanizing the irrigation work, the following savings can be achieved:

- A large number of animals useful in milk and meat production will be available.
- Irrigation water will be saved, for the reason that mechanizing the irrigation process will make it possible to provide the land with an adequate water ration according to the needs of every crop. Supposing for instance, that we should reach the ration of 5000 m^3 per feddan per year as estimated by the Ministry of Irrigation instead of 8000 m^3 being the current usage, we would save about 18 milliards m^3 of water, i.e. more than double the quantity saved for us by the High Dam. The saving of such quantities of water would provide us with the opportunity of horizontal expansion by about 2,5 million feddans of new land. It will also relieve the load of drainage for the current drainage system and enhance soil fertility. Mechanization of the irrigation operations would allow the utilization of modern irrigation methods, saving more water than the present system, i.e. injection, sprinkler and syphon irrigation, which consequently more water becoming available for horizontal expansion.

4. Mechanization and agricultural manpower

Statistics show that rates of the rural population growth are much lower than those of urban population. In Egypt the urban population increased from 36.6 % to 42.4 % and the rural population decreased from 65.4 % to 57.6 % in the period from 1950-1970. The structure of manpower during 1960-1970 showed that manpower in agriculture proportionally decreased in spite of its absolute increase from 53 % - 48.9 % of all manpower in the national economy. This is due to the economic development, urbanization and internal migration. Recent calculations in Egypt based upon the same to-day rates of economic development, population growth, urbanization and internal migration, show that the absolute number of agricultural population and manpower will begin to decrease from the year 1995. I think that the influence of programmes concerning family control and planning, the establishment of new industrial centers, the accelerated process of urbanization and the economic development will bring this phenomenon several years earlier. Our duty is to face this problem immediately. An analysis of the structure of manpower in Egyptian agriculture shows that 39.3 % of all workers are children between 6-12 years and 25,4 % between 12-18 years. They are about 45 % of all human power in agriculture. According to the new education decree all children between 6-12 years must attend school. Therefore in few years the obligatory education must cover all (100 %) of the children up to the age of 15 years. That means about 40 % of to-days human power in agriculture must be replaced by machines.

Experiments and field practices show that we need about 4 month to harvest the first crop and cultivate the next one. By mechanization of the traction processes we shall decrease this time to 2 month only. Than we can cultivate an intermediate short life crop as vegetables or animal fodder. In this case we need more manpower for harvesting these crops and that means more mechanization.

From the previous analysis we proved that mechanization is very important for Egyptian agriculture to-day and in the future, which will increase the production and raise the efficiency of the resources in Egypt.

III. Towards a plan for agricultural mechanization in Egypt

To choose the level of mechanization in order to put a plan for agricultural mechanization we need an estimation of the present tractor power in Egypt. The tractor power would replace the animal power and 40 % of human labour in order to raise the productivity of agricultural labour and to mechanize some other operations such as the insecticide and pesticide application and other operations in the animal husbandry.

The following table shows the tractor power in Egyptian agriculture with its different sources, like manpower, animals and machines.

Table 1 **Tractor units in h.p. from different sources for every farm size in 1961**

	from animals	from machines	from man-power	for every 100 feddan in every farm size group
Less than one feddan	4.70	2.50	47.20	26.20
1 feddan to less than 2 feddan	8.15	1.28	29.43	16.35
2 feddan to less than 3 feddan	9.15	1.28	26.43	14.68
3 feddan to less than 4 feddan	11.15	1.65	26.40	14.66
4 feddan to less than 5 feddan	12.55	2.02	26.57	14.86
5 feddan to less than 10 feddan	15.35	3.28	27.63	18.35
10 feddan to less than 20 feddan	20.95	10.34	38.29	21.29
20 feddan to less than 50 feddan	30.90	57.10	93.00	51.66
50 feddan to less than 100 feddan	56.35	26.84	81.99	48.88
more than 100 feddan	126.00	16.10	145.60	86.44
average for every 100 feddan	31	11.28	10.60	

This table is calculated according to the following assumptions - the traction power from different sources is calculated according to its technical power not according to its usable capacity, therefore it is more than in reality.

- all animals are estimated as traction animals, all tractors are estimated to have 50 h.p. The manpower is estimated as all workers working all the year. This makes my estimation higher than in reality.
- The statistics from the year 1961 compared with the tendency of today for the poor middle farmers shows no change; but the rich farmers over 20 feddan replaced manpower by machines in order to get more profit.

They cultivated their farms according to capitalistic type of management. If we accept 70 h.p. per 100 feddan[x] in the year 1985, it means, replacing the 31 h.p. from animals and 5 h.p. from human power, to replace the old

[x]70 h.p. is equal to the level of mechanization in G.D.R. in 1960. We can accept that one feddan needs operations under Egyptian agricultural crop rotation equal to the need of one hectare in G.D.R. and we can accept also the situation of G.D.R. in 1960 as a goal for A.R.E. in the year 1985 i.e. 25 years later.

machines, to increase the productivity of agricultural labour and to mechanize some other operations.

The following table (2) demonstrates the process of replacement and its costs. According to this table we need about L. E. 23 millions every year. The number of tractors we need every year gives good reasons to establish a national industry for farm machinery. We can begin with a multi-purpose machine like a tractor with its attached equipments such as ploughs, levellers, planters, pumps, etc.

From the national point of view we want to answer the question, who want the mechanization, who needs it, how we execute this process, how we shall face the many economic, social and political problems. In Egyptian agriculture the rich farmers (owning more than 20 feddan) have their own machines and are able to cover the mechanization by 1985. The poor, little and middle farmers (owning less than 20 feddan) have little traction power, being far from mechanization.

65 % of the entired cultivated area suffer from the absence of traction power to carry out farm operations in an accurate manner and at the proper time. Farmers have no money to buy tractors with attached equipment and credit systems cannot assist them to buy their own machines.

In order to mechanize Egyptian agriculture attention must be paid to these farmers. The cooperation in agricultural production connected with machinery stations is the only solution to mechanize 65 % of the Egyptian agriculture and to increase the productivity.

Table 2 50 h.p. tractors needed to mechanize the traction process till the year 1985 in Egyptian agriculture and the investments

Year	tractors at the beginning of the year (number)	annual additional tractors to increase the mechanization	annual additional tractors as substitution for 8% depreciation	total number of additional tractors	tractors at the end of the year	Investments in 1000 L. E.
1975	24 000	6 000	1 920	7 920	30 000	17 424
1976	30 000	6 000	2 400	8 400	36 000	18 480
1977	36 000	6 000	2 880	8 880	42 000	19 536
1978	42 000	6 000	3 380	9 360	48 000	20 593
1979	48 000	6 000	3 840	9 840	54 000	21 648
1980	54 000	6 000	4 320	10 320	60 000	22 704
1981	60 000	6 000	4 800	10 800	66 000	23 760
1982	66 000	6 000	5 280	11 230	72 000	24 816
1983	72 000	6 000	5 760	11 760	78 000	25 872
1984	78 000	6 000	6 240	12 240	84 000	26 928
1985	84 000	6 000	6 720	12 720	90 000	27 984
						249 745

The investment for every tractor and incl. equipment costs, amounts to L. E. 2 200.
The cultivated area will increase by 50 000 feddan yearly (all the assumptions and estimations are by the author).

The Present Stage and Prospects of Agricultural Mechanization in Nigeria

by S. A. Oni

I. Introduction

In its most recent usage agricultural mechanization has become almost synonymous with the use of engines, especially tractors for farm operations. However, a broader meaning of agricultural mechanization encompasses the use of tools, implements and machines for the ultimate purpose of raising agricultural productivity. Thus, the use of hand tools, hand-pushed or combine harvesters are all expressions of mechanization at different levels of technology (5).

Despite the economic, educational and other problems associated with agricultural mechanization in the tropics, many policy makers still prescribe mechanization as the most important strategy for rapid agricultural development (2).

In fact, a review of the African literature on mechanization reveals that there is no unanimity as to the appropriate level of mechanization to be adopted. While some people talk of "Premature mechanization" or the "Paradox of mechanization in labour abundant economies", others sloganize mechanization as a Panacea for all the ills of African agriculture" (3).

Regardless of this divergent policy views on mechanization, the principal goal of mechanization needs to be examined at the onset of this paper. This is particularly important since the goals may differ with differences in regional resource andowments. In general, the goals of mechanization can be summarized as follows:

"A substitution of machine power for human muscle for the primary purpose of increasing aggregate output of crops and/or livestock" (4).

Such a substitution invariably entails better efficiency of labour utilization (i.e. higher output per unit labour used). Where labour is scarce and expensive, mechanization is practized primarily as a labour-saving technique. This is true of developed countries like U.S.A. and Britain. Mechanization can also be practized for the primary purpose of increasing the efficiency of land utilization through higher yields per acre or hectare. This is sometimes referred to as land-saving mechanization and it is particularly useful in land-scarce economies.

Perhaps, a more pertinent goal of mechanization particularly from the viewpoint of social welfare is that of minimization of drudgery or irksomeness of labour. This is true in a labour-surplus economy, and even truer in a tropical environment where the enervating heat of the sun is a serious constraint on farmers' labour productivity.

It should be admitted that these goals of a mechanization are not mutually exclusive. In fact, in most tropical African countries, mechanization is geared towards the simultaneous attainment of two or more of these goals. We will return to this point later, but, in the meantime let us consider some preliminary information on Nigeria.

II. Background information on Nigeria

With a population of about 70 million people Nigeria is the largest country in Tropical Africa. It has an area of about 356 000 square miles and it is located in West Africa between longitudes 3° and 14° and latitudes 4° and 14° north of the equator. The climate is tropical with temperature ranging from 80 to 90°F all the year round.

The country is largely agricultural since about 70 per cent of the population is engaged in agriculture. The country is endowed with immense natural resources and despite the recent production of petroleum agricultural contribution to gross national product is still significant.

III. Mechanization of Nigerian agriculture - the present stage

Mechanization of Nigerian agriculture can be discussed under four major sub-headings; (a) hand tool techniques; (b) bullock or animal drawn implements; (c) tractors or tractor hiring units and (d) mechanized agricultural cooperative settlement schemes. Let us briefly examine each type and level of mechanization:

a) Hand tool techniques

A vast majority of Nigerian farmers still rely on traditional hand tools like cutlasses, hoes, sickles and axes for cultivating their farm lands. Although there is a wide range of local variation in the size of these tools due to differences in soil and other environmental conditions, the basic design of the tools are the same. They are generally made from scrap metals by village blacksmith. Except for minor adjustments in shapes and sizes there has been no concerted effort to improve these local hand tools. Because of the limited scope of these tools the average Nigerian farmer can hardly cultivate more than about 3 acres of crop land. The ultimate result of this is low farm productivity and very little marketable surplus for each farmer. In fact, the basic problem facing the Nigerian peasant cultivators today is that of low level of technology which is amplified by the use of these traditional tools (11).

b) Use of bullock or animal drawn implements

The use of cattle as a source of power was first demonstrated in Northern Nigeria in 1922 (9). From this demonstration it was apparent that mixed farming could be a highly profitable agriculture particularly in the savannah zone of the country.

In 1933 a general extension programme was launched with the primary purpose of accelerating the use of bullock as an integral part of farming in Northern Nigeria. To enhance this programme loans for the purchase of cattle and some small ploughs were advanced through the native authority. To be eligible for the loan, a farmer must have at least 20 acres of contiguous farm land, be prepared to build cattle pens, and also ready to pay back the loans with the 5 per cent interest rate at the appropriate time (8). This extension programme generated a lot of interest particularly amidst the farmers in Sokoto, Katsina and Kano areas. In fact, while there were approximately 1960 mixed farms in Northern Nigeria 1939, these mixed farms have risen to about 51,000 by 1968, and it is expected to rise to approximately 111,000 in 1980.

This impressive growth in the use of bullock power may not present a true picture of the use of bullock since some problems actually bedeviled the programme. For instance, the minimum of 20 acres of contiguous land required by the government was somehow difficult to meet. The traditional land tenure system which perpetuates fragmentation of holdings further complicates this problem. The net result is that in certain parts of the Northern States (Katsina and Kano areas) this rule had to be relaxed since less than fifty per cent of the farmers who needed assistance were unqualified by this criterion.

A more disturbing feature of the programme is however its use of only one tool - the Ransome Emcot Ridger. This tool is used for ploughing, for annual ridging, for weed control in furrow, and with the wings removed, it is used as a time to harvest groundnut. There is need to develop or improve on this implement and at present some research works are going on at Samuru to develop new improved implements that will use bullock power more efficiently.

Despite the crude implements and the inherent institutional problems, there appear to be some apparent economic of financial gains derivable from a change of use of hand tools to the use of bullock-powered implements. This economic advantage may well account for the growing popularity of mixed farming in the country as shown in table 1. However, detail discussion of the benefit-costs of bullock power use will be reserved for section four of this paper.

Table 1 Growth of mixed farms in Nigeria

Selected year	Number of mixed farms
1939	1,959
1945	2,400
1956	15,000
1968	51,000
1980	111,000[x]

[x]= Projected

Source: Federal Department of Agriculture, Lagos, 1973.

c) Farm tractors and tractor hiring services

The use of agricultural tractors is not a recent phenomenon. In fact, several tractors were imported into Nigeria prior to the second world war (13). Shortly after this war, attempts were made to introduce mechanized farming involving the use of tractors and other equipments at a commercial scale in the country. A typical example of such attempts is the famous Niger project (1). This was a large scale settlement scheme located at Mokwa about 150 miles north of Ibadan. The project was initiated in 1949 with the primary aim of producing groundnut and guinea corn from about 32,000 acres of land utilizing a high degree of mechanization. The project was liquidated in 1954 and BALDWIN (1) in his excellent book has rightly summarized the failure as follows: "We may sum up by saying that the history of the project at Mokwa is a sharp lesson that successful economic development depends on adequate knowledge. The project, even though a commercial failure, has been valuable in that it has shown that many questions must be asked before the establishment of similar kinds of mechanized agricultural and settlement schemes."

Despite this adverse experience various other forms of mechanized agriculture were subsequently tried in Nigeria. Examples include the upper Ogun estate, the Sokoto Rima valley project, the Jere Rice Bowl, and the Uzo-Uwani schemes.

The last three schemes were concerned with commercial rice production in the country. The phenomenal interest in rice might be explained by the fact that it is difficult to hand cultivate particularly during the early rains when the fadama (or swamp) is flooded with water. The use of irrigation and the possibility of two crops per annum also favour mechanical rice cultivation. Most of these rice schemes have met with only modest success.[x]

Despite these specific government schemes some cooperatives and private individuals have combined to purchase tractors for hire to the general public. The success of such cooperatives has encouraged some State governments to embark on tractor hiring schemes. The popularity of this scheme has accelerated the importation of tractors into the country as it is shown in table 2.

Table 2 Importation of tractors into Nigeria

Year	Massey-Ferguson	David Brown	Ford	All Types
	No	No	No	No
1961	72	54	67	193
1962	56	13	2	71
1963	80	75	9	164
1964	161	80	74	315
1965	85	25	29	139
1966	142	45	50	237
Total	596	292	231	1,119
Percentage	53	26	21	100

Source: Purvis, M.J.: A study of economics of tractor use in Oyo divisions of Western Nigeria. CSNRD No. 17, 1968.

The table reveals that within a brief span of 5 years the number of tractors imported into the country rose from less than 200 to a cumulative figure of over 1,000 in 1966. In fact more recent communication with the principal importers reveals that the current figure is now about 2,000 even though some of the tractors were received under external assistance programmes.

The tractor hiring services of the government are operated on a State basis. In a recent study on the economics of tractor hiring services in

[x]For instance the Sokoto Rima valley project has been liquidated although a new effort is currently under way to reactivate the project.

Western State of Nigeria by Kolawole (6) and Purvis (13) it was shown that the service was of little financial advantage to the peasant farmer. In addition, the cost of service was very high due to shortage of operators low hours of tractor use per year and some occasional breakdown of tractors and equipment. Added to this was the difficulty of obtaining spare parts for the tractors and other associated equipments.

Another significant feature of the scheme is that the hiring service is highly subsidized. In fact, most of the government charges are far below the cost of operations, depreciation and other incidental costs. These costs are approximately double what the government changes the user-farmers. Tables 3 and 4 have been constructed to portray the detail situation.

Table 3 Total cost per hour per tractor for tractor hire units in Western Nigeria

Expense item	Cost per hour per tractor
	(N1 = $1.5 U.S.)
Operating costs [(a)]	N1.44
Depreciation on equipment [(b)]	N1.15
Depreciation on building and interest on inventory of spare parts	N0.03
Supervision	N0.34
Total	N2.96

(a) Made up of wages, fuel and oil cost, as well as repairs costs
(b) See Appendix 1 for the detail.

Source: Purvis op. cit. pp. 25.

Table 4 Cost per hour of tractor operation and government charges, tractor hire units. Western State Nigeria (1965-67)

Operation	Acres per hour per tractor	Total cost per hour (a)	Government charges per hour (b)	Gain (or Loss) per hour
		N	N	N
Ploughing	0.59	2.96	2.21	(0.75)
Harrowing	1.16	2.96	2.03	(0.93)
Ridging	1.02	2.96	2.04	(0.92)
Cultivation	1.16	2.96	2.06	(0.90)
Planting	1.54	2.96	2.69	(0.27)
Fertilizer	2.02	2.96	3.03	0.07
Transport	-	2.96	1.50	(1.46)
Shelling	-	2.96	1.25	(1.71)

(a) Average operating cost and overhead costs but excluding interest on capital.

(b) Calculated from rates per acre for field operation and acres per hour for government operations.

Source: Purvis, M. J. op. cit. pp. 26.

Table 5 Types of farm work done by tractor hiring units of Western Nigeria (1967)

Operation	Percentage of total work hours
Ploughing	29.9
Harrowing/Discing	21.7
Ridging (a)	1.4
Cultivation	2.0
Planting, mowing, etc.	9.4
Fertilizer	2.2
Shelling, cutting	5.9
Transporting	25.3
Other (b)	2.1
All	100.0 (c)

(a) Probably understated and may be included with harrowing.

(b) Bunding.

(c) Does not add to 100 percent due to rounding.

Source: Purvis, M. J.: op. cit.

A closer look at both tables 3 and 4 reveals that it costs about N3 or approximately $4.50 U.S. dollars to carry out an operation with a tractor for each hour. However, the government charges depend on the specified farm operation. In fact, in most cases these charges are for below the cost since the government subsidizes the tractor hiring service. The subsidy varies from about 100 % to about 40 % depending on the farm operation.

In order to gain a better perspective of the popularity of various farm operations which are mechanized by the private farmers, we can focus on the percentage distribution of the types of work carried by the tractor hiring units. Table 5 summarizes the results for the Western State.

The table reveals that ploughing ranks first in the list with approximately 30 percent of the tractor time devoted to this activity. Transportation is second with over 25 percent whilst ridging appears rather insignificant. This might be due to the fact it is counted with harrowing.

Because of the popularity of tractor hiring service in the Northern part of Nigeria, we should discuss briefly the existing situation with respect to farming. A recent report by the FAO summarizes the situation in Table 6. Two points should be noted from table 6. The first is that charges are normally assessed on acreage basis rather than upon fixed hourly rate as in the Western State. The significant advantage of this is that the farmer is protected from indolent and inexperienced operators or even from faulty machines.

Table 6 Cost of contract services in Northern Nigeria (1962 - 65)

Operation	Actual cost	Prices paid by Farmers
A. Normal upland soils	N	N
Ploughing	3.60	1.80
Ridging	2.80	1.40
Direct ridge splitting	4.40	2.20
B. Black cotton soil vertisols		
Ploughing	6.00	3.00
Harrowing	4.00	2.00
Ridging	3.00	1.50

Source: FAO: Agricultural development in Nigeria 1965 - 1980.

The second point is that the farmer is required to have stumped his land before the tractor services are called in to carry out actual cultivation. This protexts the service from damages that may be inherent in the use of machines on a land that is not properly cleared.

An additional advantage of the service is the 50 % blanket subsidy for each farm operation. This uniformity of farm subsidy is very attractive to the farmers.

d) Mechanization at agricultural cooperative settlements

In western, Mid-West and Eastern States, the agricultural cooperative settlement programmes were launched in early 1960s. The basic aim of the programme is to encourage young school leavers to undertake farming with the use of modern machines and equipment and with a government financial assistance provided through a supervised credit scheme (10). In Western State alone there are presently about 30 settlements and on each settlement can be found an average of about 50 to 60 settlers. Although each settler has his own farm units, there is cooperative input purchasing and cooperative marketing of farm output. Tractors and other equipment are used on the settlement for the cultivation of crops. The state governments supply the tractor services at cost to the settlers. Each settler is debited with the cost of cultivation and his account is balanced after harvest. The charges levied in Western Nigeria farm settlements are shown in Table 7.

In discussing Table 7 it should be noted that the calculations for the normal field are based on the tractor-operating cost of ₦1.80 per hour. These costs are increased for small field as can be shown in column 4 of table 7.

The cost for normal field compares less favourably with the subsidized cost paid by the private farmers in Northern Nigeria is previously shown in table 6.

Table 7 Cost of contract mechanization in Western Nigerian farm settlements

Operation	Acreage per hour (acres)	Cost per acre/ normal field ₦	Cost per acre/ small field ₦
Ploughing	0.5	3.60	4.30
Discing	1.5	1.20	1.50
Ridging	1.5	1.20	1.50
Planting	2.0	0.90	1.10
Inter-row work	1.5	1.20	1.50

Source: FAO: The Farm settlement scheme of Western Nigeria, Report No. 1720.

IV. Economics of agricultural mechanization in Nigeria

The apparent dominance of traditional tools such as cutlasses and hoes, has raised the question of economic profitability of the improved level of mechanization. The question can be posed simply as thus: Are there economic benefits to be derived from the use of bullocks and tractors by the peasant farmers? If such benefits are obtainable, what are the bottlenecks militating against rapid adoption of these technologies? Let us answer these questions one at a time.

In a recent survey by C.K. Laurent (7) it was shown that there are some economic benefits accruing to farmers who use bollock in Northern Nigeria. This is because the use of bullock increases the farmers capacity as is clearly shown in Table 8. The table reveals that on the average the capacity of the farmer can be increased by as much as 300 per cent if he changes from hand tools to bullock.

In terms of acreage the farmer with a pair of bullocks could farm 12 to 15 acres. This compares favourably with the average size of 4 acres tilled by the farmer and his family of three children with only traditional tools. Hence one can say that the use of bullock triples the amount of land that can be farmed.

Also in terms of cost advantage, assuming a farmer with a bullock the cost structure can be summarized in table 9 below:

Table 8 Capacity increases: Bullock Vs. hand tillage

Estimate of the amount of land that could be farmed compared with hand tillage	Farmers					
	Users		Non-Users		Total	
	NO	%	NO	%	NO	%
Three times as much	76	69	53	31	129	46
Twice as much	25	23	84	49	109	39
Slightly more		2	17	10	19	7
About the same	2	2	7	4	9	3
Don't know	4	4	10	6	14	5
Total	109	100	171	100	280	100

Source: Summary of survey of 4 villages in Sokoto Province, North Western State of Nigeria (March - April 1968).

Table 9 Cost of farming with bullock in Northern Nigeria

Item	Annual cost
	N
Pair of bullocks (a)	6.68
Plough (at N42, eight year linear depreciation)	5.25
Yoke (at N1, four year linear depreciation)	0.25
Chain (at N3, eight year linear depreciation)	0.39
Plough points (4 in six years at N2.40 each)	1.60
Interest (at 10 per cent per annum)	1.43
Total	15.60

Source: Laurent, C.K.: op. cit. p. 245.

The table reveals that the annual added cost of operation with the use of bullock is estimated as N15.78. Assuming the farmer ploughs 15 acres per year, the annual bullock cost of ploughing an acre can be estimated as N1.05. When added to the wage cost of about N0.70 per day and assuming he can plough two acres per day the cost of ploughing with bullock is estimated as N1.40 per acre.

When one compares this cost with the cost of ridging by tractors which is estimated as N3.60 per acre (13), it becomes quite obvious that it is much cheaper to plough with bullocks even though some engineers content that tractors normally plough better (12).

Perhaps more significant is the comparison of this cost ridging by bullock with the cost incurred when traditional hoes are employed. Assuming a shadow price of forty kobe per labour (opportunity cost of farm labour), the cost per acre for hand ridging can be estimated as N6.40 per acre (12). This cost is about six times that of ridging with bullocks.

With respect to tractor use, a recent survey by Purvis (13) indicated that some economic benefits accrued to the farmers who used tractor services in Western Nigeria.

Because of availability of tractors some group farms and private individual farmers were able to double their acreage of cash crops. Not only that, the timely operation reduced the late planting period and accelerated crop yield.

Table 10 Benefit-costs from the use of tractors in Western Nigeria (1968)

Item	Group farmers	Private farmers
	N	N
Increase in gross output per farm[a]	224.60	26.40
Cost of tractor services	78.40	30.60
Net Increase in output per farm	126.40	30.80
Net Increase in output per acre	10.00	6.60

(a) Based on a maize price of N40 per ton and estimated from increase in both yields and acreage at average yield of 1,232 lbs per acre shelled maize, and average farm size for group and private farm.

Source: Purvis, M.J.: op. cit. p. 38.

In fact, farmers were unanimous in reporting an increase in yield when they used tractors (13). It was reported that about 54 % increase in yields were obtained from all crops grown under tractor mechanization. Besides this added yield, there were added costs incured hence it is quite necessary to look at the benefit-cost relation with respect to tractor users. Table 10 summarizes the results for Western Nigeria.

The table reveals that as much as N10 per acre or N6.60 could be gained by group and private farmers if they utilize tractors instead of hand tools. This added proftability should encourage the farmers to adopt tractorization but some bottlenecks are inherent in tractor use. These bottlenecks are presented under problems of mechanization in Nigeria (Section V).

V. Problems of agricultural mechanization in Nigeria

Despite the apparent profitability of the use of bullocks and tractors most Nigerian farmers still rely heavily on the use of traditional tools. A multiplicity of socio-economic and technological factors account for this situation.

Firstly, the use of either bullocks or tractors requires substantial capital investment which only few Nigerian farmers can afford. For instance, a farmer requires about N150 to start a bullock-operated farm. This is the basic cost of a pair of bullocks and a small Emcot plough. In a capital-scarce economy as ours, such an investment may not be easy for the peasants to undertake. This is why the government loans are required in order to expand the bullock programme. With respect to tractors the initial investment appears highly prohibitive and except for group farms or estates, it is highly improbable that the farmers will purchase them. A second problem of mechanization is that of low number of trained staff. The country is in short supply of agricultural engineers who can effectively operate and manage mechanized agricultural projects. This inadequacy of personnel, by and large, limits the number of tractors and other equipments that could be used in Nigeria.

Closely related to the personnel problem is the fact that most tractors and other equipment imported into Nigeria are those designed for mechanization in the Western World. In fact, a brief lock at table 2 will confirm that more than 60 % of the tractors were imported from U.S.A. The problem with these tractors and other equipments is that they are not really designed for tropical climate. While most of them frequently break down, those that work have shorter economic lifespans. Some even lie idle for over six months since the spare parts are not readily obtained in Nigeria.

A more limiting factor, particularly with respect to the use of bullock, is the presence of tse-tse flies in certain parts of Nigeria. The entire middle belt zone, which covers about 15 % of the total area of the country is unsuitable for bullock use because of the presence of tse-tse flies which cause trypanosomiasis (8) in livestock.

Most livestock are highly susceptible to this disease and there is a need for fly eradication programme in this area before bullock farming can be developed.

An additional problem of mechanization in Nigeria is the inherent land tenure system which tends to encourage fragmentation of farm holdings. While the average farmer may have about 2 or 3 farm holdings scattered about due to the bush fallow system and heriditary nature of farm ownership the successful operation of mechanized agriculture requires contiguous farm holdings. In fact, LUNING estimated that a farmer needs about 20 acres of farm land in order to successfully utilize a pair of bullocks. Not many peasant farmers can meet these requirements.

Some of the equipments used are generally ill-designed for the tropics. Those that are made by the local blacksmiths are not case-hardened hence they wear out rapidly. A typical example of this is the plough share used by the bullocks in Northern Nigeria.

Lastly there is a general problem with high mechanization due to the inherent displacement of labour. This is not a (major) problem in Nigeria at present.

VI. Policy recommendations and future prospects

Mechanization can be of immense advantage in accelerating Nigerian agricultural development. Besides the obvious economic advantages of higher productivity for crops and livestock, mechanization can really alleviate the drudgery that is traditionally associated with farm. Such an alleviation can change the attitude of the Nigerian youths to farming and thus reduce the rural-urban migration.

Some strategic policies have to be vigorously pursued in order to enhance appropriate mechanization in Nigeria. In fact, while we should realize that the country cannot afford complete tractorization of all farms at this period of our development; there is a need for the encouragement of bullock farming, particularly in the Sudan savannah zone which covers about one third of the total area of Nigeria. Both the State and Federal governments should encourage the use of bullock by providing loans to needy farmers to purchase bullock and the associated equipment, and by ensuring a simultaneous encouragement of the use of fertilizers and improved seedlings. The State extension workers will be most useful in this bullock programme. Under the present development plan, over 4 million is earmarked for mixed farms. Since this can provide loan only to about 50,000 teams of bullock together with their ploughs, there is an urgent need for revision.

Furthermore, there is an increased need for government to embark on a training programme whereby agricultural engineers and other machinists are trained to handle equipment. Presently, the Federal government is doing something positive in this direction. Other state governments should however provide training for some qualified Nigerian in order to meet up with the expanded mechanization programme under the present development plan.

Closely allied with the training of personnel is the need to develop new equipment which are designed for tropical world. This should primarily be small scale equipment which are animal-powered or those with limited horse-power. The Department of Agricultural Engineering at our universities should be given research grant to work on this topic.

Added to this suggestion, is the need to indigenize the Nigerian farm machinery industry. This is already taking root with the Nigerian enterprise promotion decree which came into effect in April this year.

Lastly, but not the least, is the urgent need for educating the masses of the peasantry about the basic economic advantages of agricultural mechanization. Since most peasant farmers are rational, once they perceive the gains derivable from mechanization, they will be in position to adjust their production technique. In this regard the role of Extension division of the various ministries of agriculture cannot be overemphasized. I will conclude this paper by indicating that the future prospects of mechanization are fairly bright even though some major problems have to be solved. It will pay the country to be cautious and systematic since rapid mechanization can itself create unemployment problems. The increased productivity has to be matched with favourable producer price and income policies in order to enhance mechanization among farmers.

The Nigerian case has shown that there are some economic gains derived from use of bullocks and tractors. Despite these apparent gains, there is a need to understand the sociology as well as the environmental conditions before embarking on a massive drive to promote mechanization among Nigerian farmers.

References

Baldwin, K.D.S., The Niger Agricultural Project; Oxford University Press, 1957.

Buchele, W.F., No starving Billions: The Role of Agricultural Engineering in Economic Development. An Inaugural Lecture at the University of Ghana, Legon, Accra, May 9, 1969, Ghana University Press, 1969.

Deutsch, A.E., Tractor Dilemma for Developing Countries; World crops XXIV, 5 September - October, 1972, pp. 234 - 236.

Ellis, G., Man or Machine; Beast or Burden: A case study of the Economics of Agricultural Mechanization in Ada District, Ethiopia. Unpublished Ph. D. dissertation University of Tennessee, June 1972.

FAO, Agricultural Development in Nigeria 1965 - 1980, Rome 1966.

Kolawole, M.I., An Economic Study of Tractor Contracting Operations in Western Nigeria. Unpublished Ph.D. thesis, Cornell University 1972.

Laurent, C.K., The Use of Bullocks for Power on Farms in Northern Nigeria. Bulletin of Rural Economics and Sociology. Vol. 13, No. 2, 1968.

Ministry of Agriculture, Langtang Mixed Farming Scheme; Newsletter No. 14, Kaduna, Nigeria, 1959.

Newly, W.T., A study of Costs of Mechanized Agriculture in Nigeria. NISER, Ibadan, 1955.

Oni, S.A., Credit in Rural Development: An Appraisal of Supervised Credit Scheme for Farm Settlers in Western Nigeria. Proceedings of the 1972 Annual Conference of Nigerian Economic Society, pp. 101-205.

Oni, S.A., Increased Food Production through Agricultural Innovations in Nigeria; West African Journal of Agric. Economics, Vol. 1, No. 1, 1972.

Philips, T.A., An Agricultural Note Book with Special Reference to Nigeria. Longmans of Nigeria Ltd., 1964.

Purvis, M.J., A Study of the Economics of Tractor Use in Oyo Division of the Western State CSNRD - 17. September 1968.

Appendix

Estimated Depreciation for a "Tractor Unit"
Western State Nigeria 1968

Item	Initial cost[a]	Life	Cost per year	Cost per hour
	N	Years	N	N[g]
1 Tractor[b]	2,200	6	366.6	0.69
1 Disc Mough[c]	380	4	95.0	0.18
1 Ridger[d]	390	4	97.6	0.18
1 Tractor[e]	126	4	31.6	0.06
Other[f]	100	4	25.0	0.05
Total	3,196	-	615.8	1.16

[a] List price less 10 per cent discount

[b] Average price of 4 major models in use in Western Nigeria

[c] Average of 2- and 3-disc ploughs

[d] 3-disc

[e] Assuming 1 trailer (hydraulic tipper) shared by 4 tractors

[f] Draw bar links, belts, etc. (estimate)

[g] Based on 535 tractor hours per year

Source: Purvis, M. J.: A Study of the Economics of Tractor Use in Oyo Division of the Western State. CSNRD - 17, 1968.

Present State and Development of Agricultural Mechanization in the Somali Democratic Republic

by Mohamed Abokur Haji

Agriculture is the basis of the economy of the Somali Democratic Republic, which is one of the less developed countries of Africa. The total area of the republic is 63.8 million ha. The largest proportion of this area is utilized for livestock production by the nomads, whose total livestock population exceeds 20 million heads of sheeps, goats, camels and cattle. Of the above mentioned area about 8 million ha are suitable for agricultural crop production, so that agricultural production can be extended since only one eighth of the total cultivable agricultural land is under cultivation at present. Most of the cultivable area is situated in regions characterized by rainfed crop production, where the average size of a farm is 1 - 2 ha. The standard of farming in these areas is extremely low, and the main tool for crop production is the hand hoe or yambo which has remained unchanged over the years. In those rainfed areas we find most of the true working farmers who is assisted by the Revolutionary Somali Government to improve farming techniques by the introduction of improved hand tools, training in the use of animal-drawn implements, and gradual introduction of agricultural machinery.

On the other hand, there is the rich and fertile irrigable land located in the inter-river areas between the Shabelli and Juba rivers. In this area there are mainly the privately owned large banana plantations and an increasing number of large state farms. Crops grown in this area are: banana, citrus, mango, papaya, vegetables, sesame, cotton, maize, etc. State farms, large banana plantations and some other big farms are employing increasingly up-to-date technologies based on mechanized work, thus constituting a sharp difference in social, technical and economic aspects compared with the small farms of the working peasants. In order to increase the area under crop production and reach self-sufficiency in food production, agricultural mechanization must gradually be introduced into the S.D.R. mainly because:

(1) land or bush clearance, levelling, canalization etc. must be carried out mainly with machinery if we want to exploit our big land resource.

(2) a vertical increase of agricultural production is only possible with the introduction of mechanization as proved by the stage of development reached by the large plantations in our country.

(3) the presence of tse-tse flies (or Tripanosomiasis) in most of the irrigable areas inhibits and excludes the keeping of draught animals in the vicinity of river areas.

(4) in severally eroded agricultural lands of the Northern regions, the use of farm machinery for the construction of soil bunds is the best solution for protecting fertile agricultural lands from damages caused by water erosion.

(5) the agrarian policy of our Revolutionary Government is based on the establishment of agricultural cooperatives and state farms as the basis of the future socialist pattern of economy of the Somali Democratic Republic, and this goal can only achieved by gradual mechanization of agricultural production.

For these reasons, the introduction of mechanization into the agriculture of the S.D.R., especially for state farms, was started almost 12 years ago, when an economic agreement was concluded with the Soviet Union, and a number of tractors and implements were imported from the Soviet Union. Then the National Pool of Tractor Organization (ONAT) was established.

ONAT is a state organization responsible for renting farm machinery and implements to individual farmers, cooperatives and state farms.

Recently, ONAT had 367 tractors, out of which 121 were of the type S-100 and the rest various types of light powered tractors. Compared to

1970, the tractor output (in hours) in 1971 was increased by 16.8 % and by 33.5 % in 1972. In 1973 an increase of 71.4 % was envisaged compared to 1970.

It should be noted, however, that the largest increase was obtained by the better utilization of the 100 h.p. tractors. In addition, there is also a relatively large number of privately owned farm machinery. In spite of a greater level of utilization, the agricultural policy to increase crop production led to a shortage in the quantity of farm machinery. In order to meet this increasing demand, ONAT has recently imported farm machinery from friendly socialist countries. Since irrigated land is to be vigorously developed during the 1974-1978 programme period, heavy duty machinery for such a work as land reclamation, land levelling and opening up of irrigation networks will be needed equally additional tractors are required for land utilization. ONAT is also building repair shops and garages in a number of agricultural districts at a cost of about 1,551,574 Somali Shillings.[+] In the past, ONAT was mainly used by big banana plantations and state farms, but in the forthcoming period and after promulgation of Law No. 40 on the cooperative development in the Somali Democratic Republic of October 4, 1973, ONAT will be an important factor in agricultural cooperative development by:

(1) carrying out heavy and specialized operation in land reclamation including irrigation networks in contract with state farms and cooperatives. The extension of these capacities is necessary in order to meet the increasing requirements.

(2) carrying out agricultural mechanized operations based on contracts with cooperatives and individual farmers. Therefore the network of stations and working groups should be extended. In order to use the capacity of the tractors and machinery more effectively, shift work must be introduced on a large scale.

(3) purchase and sale of tractor machinery and spare parts according to the requirements of the whole agriculture in conformity with the plan of

---

[x](1 So.Sh. = 0.40 M)

mechanization approved by the Ministry of Agriculture

(4) assistance to agricultural cooperatives in introducing modern agricultural technologies based on mechanized and semi-mechanized work, combined with employment of draught animals, especially in rainfed areas.

The main task of ONAT is to promote the cooperative development in every village, and to assist the working farmers by technical, cultural and political means. Therefore ONAT is not a profit-oriented organization but a genuine service. It applies social-economic tariffs for operations to be rendered.

ONAT should also extend its activities from field preparation to the hiring of pumping sets to the farmers, since this is an important condition to obtain stable and increased yields. The possibilities are being discussed for improving the system of tractor hiring by ONAT to agricultural cooperatives and state farms based on hectares instead of tractor hours. The assistance to the farmers with modern machinery must be supplemented with improved seeds, insecticides, extension work etc. from the Agricultural Development Cooperation (ADC) to individual farms and cooperatives.

An effective way to promote modern technologies of production is the link-up of agricultural producers to the processing industries.

In our country, raw cotton has to be imported to feed the Somaltex textile factory, since no adequate quantities of cotton are produced in Somalia, although the resources are available. The best solution for this problem has been to select suitable agricultural zones for the production of cotton, tobacco etc. to feed local factories with domestic raw materials. These agricultural producers, either state farms or agricultural cooperatives, will conclude contracts for the production, supply and mutual assistance with the processing factories. These are the first steps towards agri-industrial integration in the S. D. R. as an example for the promotion of the whole agriculture. For the gradual transition of our agriculture to a socialist agrarian system, the establishment of state farms is proceeding progressively. State farms are also needed as pilot schemes of modern large-scale production, to be followed by cooperatives. They are to be the centres of training, research, breeding and testing of new agricultural technologies.

For this reason, several state farms have been established of which Fawole, Balad, Mordile, Jowhar, and the agricultural crash programme are some examples.

The agricultural crash programme started in early 1970 has passed its initial stage of development. Its primary objectives can be summarized as follows:

(1) To increase national crop production to attain sustained self-sufficiency, with emphasis on import-substituting crops.

(2) To minimize migration from rural areas by offering employment to active pioneers.

(3) To reorient the occupational attitude of the Somali nomads towards the farming occupation by improving their skills.

Technically, the scheme concentrates on the mechanized production of agricultural commodities which are in short supply in the country. At present, the scheme cultivates a total area of 7 550 ha, of which 4 050 ha are in the irrigated areas and the remaining 3 500 ha in the rainfed areas. The land is cultivated mainly with cereals such as sorghum and wheat.

We are now entering a new stage of development by establishing agricultural cooperatives based on the principles of scientific socialism. This is of paramount importance for the introduction, popularization and development of farm machinery, because up to now the individual working farmers never had the possibility to improve and mechanize their farming technique for lack of knowledge, capital and other social, cultural and economic factors.

Since the concept of developing agricultural cooperatives under a socialist system is new to our farming communities, the law provides a period of assimilation during which each farmer will retain his holding but will be a member of a multi-purpose cooperative society, which is to undertake the marketing of farm produce, supply of farm requisites, provision of credit, communal self-help tasks, and to provide technical services either with tractors owned by the cooperative society, or according to contracts with ONAT. After creating consciousness and mutual trust among the members of the multi-purpose cooperative organizations, the next step will be the formation of group farming cooperatives, the aim of which will be to reclaim and cultivate cooperatively a certain area besides the holdings of the members, which remain with the individual farmers. The group farming cooperatives will gradually enable the farmers to understand the advantages of pooling their resources for their common interest.

The third stage of cooperative formation will be the foundation and development of cooperative farms. Here, the joined farmers and agricultural workers pool their land, means of production, and working power for cooperative production and distribution of income.

In order to achieve better farm incomes and maximum usage of land and labour, proper planning and management and modern techniques will be employed so as to demonstrate the advantages of progressive large-scale farming.

A practical example of the above mentioned strategy is the foundation of the first agricultural multi-purpose cooperative society named Kulmis Beerey at Ugunji in the Merca District. Kulmis Beerey was established on March 6, 1974, and comprises about 500 working farmers cultivating 600 ha. Members of this cooperative have for the first time introduced farm machinery. The Ministry of Agriculture has assisted them with a number of tractors, which permitted the cooperative members to increase the area under crop production and plough the soil of their plots, and this led to increased crop yields.

To sum up, the Revolutionary Government of Somalia together with the progressive forces are going ahead to achive the above mentioned goals. The foundation for achieving them is:

- revolutionary orientation, based on scientific socialism
- increasing self-help actions by the people
- utmost assistance in developing state farms and agricultural cooperative societies by all government authorities.

These are the main instruments in order to reconstruct the agriculture of the S.D.R. on a progressive basis, which will include mechanization, the integration of crop production and animal husbandry, and the gradual introduction of agro-industrialization.

Improvement of Dry Land Farming in Northern Iraq

by Abdullah K. Al-Fakhry

Introduction

Dryland farming is the main feature of agriculture in northern Iraq where 75 % of the country's total wheat and barley crops are grown (3). The total cultivated area of this region is approximately 4, 000, 000 hectares. Sheep husbandry is another equally important farming enterprise of this region and is largely based on natural pastures. The crop production per unit area is very low partly due to insufficiency and irregularity of rainfall on the one hand and lack of the technical knowhow about dryland farming on the other.

Everywhere in the region crop-fallow rotation is followed by the farmers to conserve moisture but so poorly managed that it does more harm than any good (1). They strongly believe in more tillage operations at greater depths during the fallow period to obtain better yields.

The application of chemical fertilizers and herbicides is very rare and the farmers of the area consider the manure left by sheep during follow grazing, adequate for keeping up the soil fertility (3). A large portion of the straw is removed from the fields during or just after harvesting for feeding the animals and whatever is left is gradually grazed to a point where the soil is completely exposed to erosion.

The Department of Field Crops, College of Agriculture and Forestry, Mosul University in cooperation with other departments of the college has set up a comprehensive programme for finding out better farming methods and practices in order to replace the traditional ones with a view of increasing the productivity of the rain-fed zone in this part of the country.

Characteristics of rain-fed areas of northern Iraq

Semi-arid Mediterranean climate covers the major cultivated areas of northern Iraq (1). Here the rainfall comes in winter and ranges annually betweem 250-500 mm followed by completely dry summer with high evapotranspiration. In this region the precipitation not only fluctuates as to quantity and distribution within a season but also from year to year.

Table 1 below shows the distribution and intensity of rainfall recorded for the last five years at Hammam Al-Alil, which is located in this region.

Table 1 Precipitation amounts (mm) by months for Hammam Al-Alil, Iraq

Year	Oct.	Nov.	Dec.	Jan.	Feb.	Mar.	Apr.	May	Jun.	Total
1968-69	9.5	35.5	85.6	144.3	23.0	117.7	67.4	35.4	-	518.4
1969-70	16.6	23.3	61.4	143.1	7.9	59.6	8.7	0.5	-	321.1
1970-71	-	28.6	28.6	1.9	12.0	51.9	130.9	0.6	-	254.5
1971-72	1.7	41.0	54.6	49.5	77.1	60.3	128.3	31.9	5.2	449.6
1972-73	18.5	31.5	49.0	39.8	54.0	17.3	19.9	27.0	-	257.0

Winters are generally mild and the growth period begins in March and ends in May. There is a partial coincidence between the rainy season and the active growth period. The crops adapted to these conditions are of short life cycle and have shallow roots such as wheat, barley, annual medics, clovers, rape seeds, and safflower etc. The perennials are xerophytic in nature to some degree which enable them to withstand the drought conditions and make use of the available moisture.

The soils of these areas have a weak surface structure and are low in organic matter because the climate favours its decomposition rather than its accumulation (2). The stability of the soil structure is mainly dependent on the abundance of annuals and their growth and decay season by season.

The indiscriminate use of land in these areas has altered the original balance among plant species, soils and hydrology. Consequently the soils remain bare of vegetation for long duration and are exposed to vagaries of nature particularly to wind and water.

Research programme

The comprehensive research programme which has been started from the year 1971-72 by the Department of Field Crops at the College of Agriculture and Forestry, Mosul University includes all the major crop production problems of immediate attention. The entire research programme has been divided into two sub-groups namely breeding and cultural practices.

The studies under breeding programme are mainly directed towards improving the hereditary characteristics of those crops which are of special concern to the area. The targets fixed are to come up with the varieties that will withstand drought, have shorter live cycle, resist diseases and insects respond to fertilizers, and give higher yields.

Different cultural practices are being tried with the hope to achieve the two principal goals of controlling soil erosion and increasing production per unit area at a minimum cost.

All experiments were layed out at the College farm at Hammam Al-Alil where the annual precipitation is approximate 300 mm. In addition, two more locations were selected to repeat the main results of the previous studies conducted at Hammam Al-Alil. These locations are at Rabeah and Tel-Afer where the annual precipitation is 450 and 350 mm respectively. Both locations are considered among the main wheat and barley production areas in northern Iraq.

Brief description of these research activities has been given below.

Breeding programme

The breeding programme underway includes the following research projects.

1. Introduction: A number of varieties of the following crops were imported from different dryland areas of the world resembling in nature to northern Iraq.
   1. Common wheat (Triticum aestivam): A total of 245 entries were tested for the last two seasons (annual precipitation was 450, 202 mm respectively). Five new varieties proved to be superior in yield over the existing local ones. They are India 66, Australia 153, ORG 6571,

Australia 152 and Erika 1. Their yield ranged from 1500 to 1600 kg per hectare in 1972 at Hammam Al-Alil. The tests are being repeated this year at all the three locations previously mentioned for final evaluation.

2. Barley: 57 entries were involved in the varietal trials of this crop but no significant performance was recorded for any when compared with the selections from the local materials.

3. Oil seeds:

   a) Safflower (Carthamus tinctorius): Out of eight varieties under trial only one variety "Firo" yielded 1 ton/hectare when the annual precipitation was 202 mm.

   b) Rape seed (Eruca sativa) and (Brassica campastis): Variety trials with 7 entries were started in 1972-73. Two Indian varieties performed well both in yield and oil content.
   Work with oil seeds is still continuing because more information are needed on the performance of superior varieties before final evaluation is made.

2. Hybridization: Crosses between the superior introduced varieties and the leading local ones of wheat and barley were started in the year 1971-72. The segregating materials are under examination for various characteristics.

3. Individual plant selection: A broad based programme was initiated in the year 1971-72 on the major rain-fed crops of the area - wheat and barley. A new variety of barley ("Mosul 72") has already been released as a result of this programme. A few superior wheat strains are in the final stages of evaluation.

4. Survey of native pasture legumes: Although northern Iraq is rich in native annual medics, clovers and vetches, no scientific efforts were put in the past to explore the possibilities of utilizing these native pasture legumes to the best of the area. It was, however, realized in 1972-73 to start work in this direction. Fourteen ecotypes representing 8 species of the genus Medicago were collected for examination and it is hoped to release

those of agronomic value as cultivated species for commercial production.

Cultural practices

The following studies in regard to cultural practices are in progress for the time being.

1. Effect of various rotations on wheat yield: The purpose of this study is to replace the wheat-fallow rotation being practiced in the region by certain pasture legume-wheat rotation which will not only increase the soil fertility and protect it from erosion but also provide nutritive fodder for the livestock. The treatments under observations are wheat-fallow (C K), continuous wheat, wheat-legume for seeds (lentils) and wheat pasture legumes. All the rotations need to be repeated for several times on the same land to reach definite conclusions on the economical aspects of the most suitable rotation for the region.
2. Tillage operations for wheat production: The farmers of the area cultivate their fields 3 or 4 times at the depth of 15 to 25 cm for wheat production. These operations cost a great deal of money besides loss of valuable soil moisture and soil damage by erosion. The two experiments designed for finding out the optimum number of tillage operations and depth of ploughing are as follows:
   a) The effect of various number of tillage operations on wheat yield in wheat-fallow rotation: The five treatments involved in this study are four cultivations, three cultivations, two cultivations and one cultivation only during fallow period.
   b) The effect of depth of ploughing on moisture depletion from the soil as reflected on wheat yield: Three different ploughing depths of 7.5 cm (3 in.), 15 cm (6 in.), and 22.5 cm (9 in.) are included in this experiment. The preliminary results for both tillage experiments will be available at the end of this season.
3. Effect of different seeding practices on the stand and yield of wheat crop under Hammam Al-Alil conditions: Different seeding practices are presently in use in this area but the most common practice is to distribu-

te the seed in the field by hand broadcasting and cover it by animal drawn cultivator. Seeding by disk tiller-seeder is also gaining increased importance these days.

Regular work on different seeding practices was started this year (1972-73) to standardize a method most fruitful to the area.

The investigation is being carried out at two locations on the college's farm with the treatments of seeding with grain drill, seeding with disk tiller-seeder, seeding with disk tiller-seeder with delivery tubes removed, seeding by hand broadcasting and covering seed with cultivator.

The first hand data will be available for analysis at the end of the current season.

4. Determination of the optimum rate of seed for wheat crop: Five rates of seed 20, 40, 60, 80, 100 and 120 kg/hectare were used for this purpose in the year 1972-73. The rates of 60 and 80 kg/hectare showed better performance as to grain yield. The tests are being repeated this year to confirm the results of the first year.
5. Effect of phorphorus fertilizer on wheat yield: Previous work on nitrogen fertilizer has indicated no effect on the wheat crop in this area due to low annual precipitation. Similar results have been reported from other countries of similar climatic conditions (1). Initial work with phosphorus fertilizer on the college's farm showed that wheat yield was doubled when 180 kg/hectare of triple superphosphate (47 % P_2O_5) compared with no fertilization.

   The study was set up this year at the three locations consisting of five treatments of 0, 60, 120, 180 and 240 kg/hectare of superphosphate (47 % P_2O_5) for wheat, rape seed, and safflower crops.

Summary

Dryland farming is the main feature of agriculture in northern Iraq where 75 % of the country's total wheat and barley crops are grown. Sheep husbandry is also concentrated in the same belt. Crop-fallow is the main rotation being practiced in the area. The incidence of rain is only in winter and ranges between 250 and 500 mm per annum. The quantity and distribution of rainfall fluctuates from year to year.

Research programme for crop production under semi-arid conditions covering their two main aspects namely breeding and cultural practices was started in 1971-72 at the College of Agriculture, Hammam Al-Alil. These are considered to be the immediate agronomic research problems of the region. Under the breeding programme the research studies include the introduction, evaluation and selection from imported as well as local promising varieties of wheat, barley, rape seed, safflower and some pasture legumes. These efforts have resulted already in the release of a new variety of barley and also promising strains of wheat have been obtained out of the individual plant selection project. The cultural practices programme covers studies on rotations, tillage practices, seeding methods and rates and fertilization. The results of these studies so far have yielded very useful informations of economic significance.

References

(1) de Brichambout G. Perrin (1970), Similarities and Differences in World-wide Dryland Farming. Proceedings of International Conference on Mechanized Dryland Farming held in 1968 at Moline, Illinois, U.S.A. and sponsored by Deere and Company.

(2) Downes, R.G. (1970), Principles of Applying Conservation Methods in Australia. Proceedings of International Conference on Mechanized Dryland Farming held in 1968 at Moline, Illinois, U.S.A. and sponsored by Deere and Company.

(3) Tammimi, S.A. and A.K. Al-Fakhry (1973), Some observations on Crop Production in Rain-fed and Irrigated areas of northern Iraq. Publication No. 21, United Nations-FAO/University of Mosul, College of Agriculture and Forestry, Hammam Al-Alil, Mosul, Iraq.

Tasks and Results of the Testing Station at El Gasair, A.R. Egypt

by P. Glanze

This report gives an outline of the range of functions and results of the El Gasair testing station for the June 1970 to July 1973 period.

This station is jointly operated by the VVB Land- und Nahrungsgütertechnik (German Democratic Republic) and the General Organization for Utilization and Development of Reclaimed Land (Arab Republic of Egypt), and is located in the village of El Gasair in the Mariut reclaimed-land sector.

According to the contractual arrangements this station shall deal with solutions for mechanizing plant production in Egyptian agriculture using farm machinery produced in the German Democratic Republic. The work to be done in this connection includes the following duties:

- To compare variants of equipment for determining the schedule of equipments for farm machinery to be delivered by the German Democratic Republic to the Arab Republic of Egypt
- To find out optimum machine setting and economic data so that recommended values can be made available for Egyptian farmers
- To study possibility of making some modifications to conventional methods of growing certain kinds of crops, in order to facilitate, or make possible, the use of machinery
- To make demonstrations with the object of encouraging farmers to use modern methods and farm machinery.

In the period under review, efforts centered around the testing of the E 512 combine harvester as the main machine for harvesting kernel crops as well as the E 301 swath mower, the main machine for harvesting forage. Also tested within the framework of this programme were a number of other machines and implements including:

- Tillage equipment (grubbers, fine grubbers, spike tooth rollers, disk harrows)
- Drilling machines (rice, cotton, and peanut drilling)
- Fertilizer distributors
- Potato harvesters
- Beet topper loaders and beet lifter loaders
- Mobile cranes.

In addition to the testing of farm machinery, experiments were made with the object of making some modifications to conventional methods of crop growing and selecting the most suitable varieties of plants, all this being done with the most efficient use of machinery in mind and for the following types of crops:

- Corn (flat and dam seeding, seed quantity and density)
- Rice (seed rate, distance between rows, depth of sowing, and water regime for direct seeding)
- Wheat (seed rate, fertilizing)
- Safflower (stem diameter as a function of plant density)
- Cotton (preparing seed for drilling, seed rates, flat and dam seeding).

Whereas between 1967 and 1969, the initial period of operation of the testing station, the area available to the station for making practical demonstrations of modern farm machinery and for gaining practical experience had a size of approximately 100 hectares, the work done in the period under review was not only carried out at El Gasair. A good deal of the work was done in the sectors Mariut, North Tahrir, Northwest Delta, West Nubaria and Hamoul under the jurisdiction of the General Organization for Utilization and Development of Reclaimed Land, Ministry of Land Reclamation, as well as in farms in the Kafr El Sheikh region under the jurisdiction of the General Organization of Production, Ministry of Agriculture. Of course, working on an area as large as this inevitably imposed rather high requirements on preparations for projects and also made very great demands the entire team of workers at the testing station. However, this large range of action was necessary for various reasons.

First, irrigation and drainage systems as well as field sizes (the term field being used here to describe an area that may be tilled without interruption due to relocation of machinery) are, in general, entirely different in different reclaimed-land regions. For example, in the sectors of Mariut and North Tahrir fields irrigated by catchment water usually measure 100 meters by 200 meters, the field size thus normally being about 2 hectares. In rice-growing regions as well as in regions where land is reclaimed from saline soil such fields are subdivided, by additional open drain ditches, into smaller units of 0.2 to 0.3 ha, typical regions being the sectors Northwest Delta and Hamoul. In the sector of West Nubaria, on the other hand, fields measuring 350 by 1,150 meters or approximately 40 hectares, respectively, will be found. Such sizes are obtained by a syphon method of irrigation, where the water ditches for the syphons are closed prior to mechanical cultivation of the field with tractors and special tillage equipment and then re-opened for the next irrigation period.

The second reason for expanding the station's range of action was that different crops are grown in the different regions. For example, almost no cotton is grown in the Mariut sector. Therefore, experiments with, and field test of, this particular type of crop were made in the Northwest Delta sector. Concerning the kernel crops there were also considerable differences between regions, with rice being grown in the East Delta, Hamoul, Mansour, and Northwest Delta sectors and corn chiefly in the Mariut, North Tahrir, and Northwest Delta sectors. Similar differences also exist for other kernel crops. A thorough test of the combine harvester thus called for work in different regions.

Third, an attempt was made to include new imported varieties of wheat and rice in the programme of combine harvester testing. However, these new varieties were initially grown on farms operated by the Ministry of Agriculture in the Kafr El Sheikh region. Consequently, is was necessary to transport all the machinery required for this purpose to that region.

Fourth, consideration had to be given to the fact that the area of reclaimed land represents only about 10 percent of the total area under cultivation and accounts for only 1.7 percent of the country's grain area. In the long run, therefore, mechanization of traditional regions under crop is of great importance to the agricultural machine market. The work done in the Kafr El Sheikh region gave first experiences under the local conditions typical for the old Delta.

It is not possible to give a detailed description of the results obtained in these investigations in this paper.

In order to give an idea on the results of studies made in the period under review, the author gives a number of examples showing the practical testing of the combine harvester during the harvest of rice, safflower and sunflowers, which are typical for the work done at the testing station. As an additional information, it should be noted that the combine harvester was also tested in the Arab Republic of Egypt in harvesting wheat, maize, rape, flax, broad beans, peas, cow beans, soybeans, alfalfa, Egyptian clover, Persian clover and Italian rye grass.

The "rice harvest" example will now be dealt with. For the combine harvest of rice the following questions had to be answered:

(1) Is the combine's standard undercarriage suitable for use on rice soils?
(2) Is the combine capable to handle the large quantities of wet material to be threshed?
(3) Which equipment of the threshing-unit and which settings should be used in order to insure complete threshing with only a very small proportion of broken and shelled grains?
(4) What is the efficiency of combines harvesting on rice fields divided into small plots?

Regarding the first question, i.e., the suitability of the combine's standard undercarriage for use on rice fields, use of the combine harvester on various rice fields in the Northwest Delta and Hamoul sectors as well as near Kafr El Sheikh in the old Delta showed that in the case of carefully cleaned drains with a normal water run, two weeks of preharvesting drainage were sufficient to use the combine successfully. True, the soil still has a certain degree of plasticity, but the combine is satisfactorily supported by it.

Some wetter spots due to inadequate preplanting levelling must occasionally be reckoned with, so that it may sometimes be necessary to drain for three weeks prior to harvest. Also, problems may occasionally arise from erosion gullies at the water inlet or outlet points of fields, which may easily result in the combine harvester's getting stuck if the combine operator fails to see them in time. It is advisable, therefore, to make concentrated use of combines on a location of plots and to station there a caterpillar tractor for pulling out purposes.

What has been said up to now applies to the dry season. Winter rains, which usually set in late October, may make fields impassable for weeks. Dual tires and cage wheels will then turn out to be of no use. Early planting and growing early-ripening varieties on a certain part of the rice areas are important conditions for using combines equipped with standard-type wheels and avoiding the higher-cost use of track-type undercarriages.

Before answering the second question which is concerned with the combine's capability to handle large quantities of wet rice material it is appropriate to indicate that the yields of rice are rather high in the Arab Republic of Egypt, the national average being 5.3 tons per hectare. However, during the tests made in the Delta region we frequently had to cope with yields as high as 8.4 tons per hectare.

When one further considers that the Egyptian varieties have a grain-to-straw ratio of 1 : 2, that the moisture content of straw at the time of harvesting is between 50 and 70 percent, and that the straw is fiberized by the tooth-type threshing-unit used for rice threshing, which usually involves a considerable release of moisture, it is immediately becoming obvious that an extremely great strain is put on the combine harvester. Very difficult in this particular connection is the satisfactory separation of grains by the straw-walkers, for the straw-walkers are covered with a cushion of straw that may be compared with a wetted mattress filling. On fields giving yields as high as those mentioned above, lowest driving speeds were therefore used, in addition to a reduction in effective working width to 2.8 m (1/3 of the 4.2 m header). These measures enabled straw-walker losses to be reduced to less than 1 percent. More favourable conditions for combine harvesting are being created by the transition in the Arab Republic of Egypt to the growing of imported short-straw long-grain varieties of rice on 30 to 50 percent of the land under rice cultivation, as these varieties have not only a narrower grain-to-straw ratio, but also the additional advantage of non-lodging thus allowing larger heights of cut than Egyptian round-grain varieties of rice which tend to lodge.

As regards the third question which is concerned with the equipment of the threshing-unit as well as with the thresher settings to be used, it should be premised that the adaptation of the E 512 combine harvester to different threshing conditions is obtained by threshing cylinder speed regulations and by equipping the concave with a different number of spike segments. Three variants being used namely:

2 spike segments for light threshing conditions,

3 spike segments for medium threshing conditions; and

4 spike segments for heavy threshing conditions.

Tests showed that the variant using 4 spike segments in the concave of threshing-unit required a threshing cylinder speed of 500 to 540 rpm for the Egyptian varieties and 390 to 420 rpm for the new long-grain varieties of rice, in order to obtain satisfactory results.

The low speed required for long-grain rice seems to indicate that it should be possible for these varieties to be satisfactorily threshed with only two spike segments in the concave, which could bring with it a smaller degree of defibring of straw and, thus, a more satisfactory separation of grains by the straw-walkers. However, it was no chance so far to make practical tests of this particular variant.

The fourth and last question is concerned with the efficiency of combine harvesting on rice fields divided into a large number of plots. The small fields (which normally measure 25 by 80 meters), the low driving speed, and the reduced width of working result in a much lower output per unit of area than that which is obtained for wheat.

While during the wheat harvest on fields averaging two hectares in size, 1.5 hectares were harvested in one hour, the performance on rice fields was 0.2 to 0.7 hectares per hour, which is still a high efficiency considering the particular conditions.

Now some remarks to our second example, the work on safflower. Whereas, for rice, the principal task was to test the attitude of the combine harvester for various varieties at different locations, safflower harvest posed entirely different problems. It was necessary to find out how to modify the conventional method of growing this crop so as to enable the latter to be harvested by combines or, in other words, to adapt the crop to the machine. The space alloted to individual plants by traditional methods of growing is 40 to 60 by 25 to 30 cm. This results in tall and stout plants, of which the stalks have a diameter greater than 20 mm and which at the time of harvest are so lignified that the knife-blade rivets are frequently sheared off during cutting.

Growing experiments using different distances between rows and seed rates showed that, on the recultivated calciumcarbonate soil of the test area, a spacing between rows of 41.7 centimeters and a seed rate of 50 kilograms per hectare (without singling) gave stems that were thin enough for combine harvesting.

On the other hand, sunflower, our third example, required a machine-to-crop adaptation. We had to modify the combine's header so that the latter would be adaptable to sunflower harvesting. We first tried to use the corn mowing attachment where the real was lined with hessian cloth. However, losses due to seeds falling out or heads being broken off were too high. The next step was the development of an additional attachment for the standard-type header, which was designed and constructed in close cooperation with G.D.R. experts working at the central workshop of the Mariut sector. This attachment, to which several modifications were made within a two-year testing period, is now being manufacured locally.

These few examples indicate how different the tasks were alone for the successful test of the combine harvester. At the same time these examples are representing the working method of the station:

1. Testing machines under conditions determinated by the locations and the varieties of crops (example rice)

2. Adaptation of plants and growing technology to the machine (example safflower)
3. Adaptation of the machine to the plants (example sunflower).

Of course it is often necessary to combine these ways to get quickly practicable results.

The most important result of the extensive tests made in the period under review is that it was possible to show with conviction that large self-propelled machines such as the combine harvester and swath mower may be successfully and efficiently used even on irrigated land in Egypt.

Furthermore, the most suitable equipment of machines could be determined experimentally. Also resulting from extensive experimental work were recommendations for both machine setting and machine use. What was of the utmost importance in this connection was that it was possible to demonstrate the advantages of using combines for harvesting rice, since the mechanization of rice harvest may have a decisive influence on the development of the combine harvester market.

Important is now the question how the results of these investigations were evaluated and put into practice. This was done in various forms. First, a written report of each test result was submitted to our Egyptian partners. This was then followed by practical demonstrations to Egyptian experts, demonstrations being made of the harvesting of wheat, corn, and rice using combines and of hay making using the swath mower, combination hay rake and tedder, and high-pressure hay baler. Also, lectures on the results of our work were given to audiences at the University of Alexandria. Additionally, farmers were instructed in the use of modern machinery. Moreover, a leaflet printed in Arabic and giving information on the direct seeding of rice was distributed by the General Organization for Utilization and Development of Reclaimed Land. Official events such as visits of delegations, negotiations, etc. were also used to propagate the results of work done at this station. A high point in this connection was the visit to this testing station by GDR minister of agriculture and his delegation.

The consequences of the station's activities were many. The testing station made an important contribution toward strengthening the prestige of the German Democratic Republic and the reputation of its farm machinery industry demonstrating true cooperation. On the one hand it helped to solve many problems associated with the mechanization of Egyptian agriculture and on the other hand it gave valuable information to farm machinery, manufacturers and suppliers on the proper equipment of machinery to be supplied to Egypt. The station's activities also had direct commercial consequences.

For instance since the E 512 type of combine harvester was tested only combines of that type have been imported by the Arab Republic of Egypt, the same being also true of the E 301 type swath mower.

Last not least it must be stressed that the good results of the station were only possible on the basis of a good cooperation:

- within the team of the station between the GDR-experts and their Egyptian colleagues
- between the station and the Egyptian governmental administration, represented by the General Organization of Reclaimed Land Cultivation
- between the station and the farms and the reclaimed-land-sectores
- between the station and Egyptian scientific institutions.

Finally it can be said that the extensive testing of machinery by a jointly operated testing station and the instruction given to farms on the use of job-proven machines has satisfied all expectations. Our Egyptian partners repeatedly told us that this method of working together should serve as a model and that they consider it an expression of real cooperation and a valuable contribution toward strengthening the country's economy and overcoming the effects on the latter of the Israeli aggression as soon as possible.

Land Tenure and Agricultural Mechanization in Egypt

by Mohamed Abou Mandour El-Deeb

Land tenure is a broad term covering all those relationships established among men that determine their varying rights in the use of land. It deals with the splitting of property, rights, or their division among various owners, between owner and occupant, and creditor, and between owners and the public, it includes assessment of taxes on private rights, and regulation of land use through various social control devices. Land tenure refers also to the period during which rights in land are hold. (1)

The situation of land tenure in Egypt is one of the most effective factors in the agricultural mechanization. The statistical data about land ownership from 1952 - 1965 is characterized by great inequality (see table 1).

From table 1, we can see that land ownership in Egypt has changed in the direction of increasing the number of small farms less than 5 feddans in area, and in the number of owners. It is due to the laws of the land reform in the country.

These farms are too small to provide a subsistence minimum for the cultivator and his family, or to provide them with full employment, or to permit any improvement in methods of cultivation. This situation of farms caused by the pressure of population on land and of inequality in the distribution of land ownership.

The small farms are also fragmented into numerous different plots scattered over wide areas. This fragmentation causes waste of time and effort and makes rational cultivation impossible. The high cost per feddan -

Table 1 Distribution of land ownership in Egypt in 1952 and 1965

Size of farm ownership in Feddan	1952				1965			
	Land-owners		Area owend		Land-owners		Area owend	
	in thousand	% to the total	in thousand Feddan	% to the total	in thousand	% to the total	in thousand Feddan	% to the total
Under 5	2642	94.3	2122	35,5	3033	94,9	3693	57,1
5 to 10	79	2,8	526	8,8	78	2,2	614	9,5
10 to 20	47	1.7	638	10,7	61	1,8	527	8,2
20 to 50	22	0,8	654	10,9	29	0,8	815	12,6
50 to 100	6	0,2	430	7.2	6	0,2	392	6,1
100 to 200	3	0,1	437	7,3	4	0,1	421	6,5
Over 200	2	0,1	1177	19,7	-	-	-	-
	2801	100	5984	100	3211	100	6462	100

Source: Ministry of Agriculture, Department of Agricultural Economics and Statistics, Monthly report - 1968 pp. 11-14 (in Arabic).

of surveying and exchanging many small plots is a major obstacle, as well the conservation of the peasants (see table 2).

Table 2 Small properties fragmentation in 1950 and 1961

Property structure	1950 Number of plots	1950 Average size of plot in Fedd.	1961 Number of plots	1961 Average size of plot in Fedd.
Less than 1 Feddan	1.28	0.41	1.50	0.32
1 to 2 Feddans	1.87	0.72	2.27	0.58
2 to 3 Feddans	2.34	0.99	2.80	0.81
3 to 4 Feddans	2.72	1.22	3.14	1.03
4 to 5 Feddans	2.92	1.47	3.44	1.23
Average	2.00	0.91	2.33	0.73

Source: Ministry of Agriculture, Statistics and Census Department, Agricultural Census for 1950, and for 1961 (in Arabic).

From the table we can see that the small farms in Egypt are going towards more fragmentation which tends to further decrease of the average scale of the plots.

This structure of land ownership effects the possibility of agricultural mechanization. The big farms can hold tractors and other farm machinery for many reasons:

(1) the scale of the farm allows full employment for these machines and other factors of production;

(2) the big farms give more profits from the unit of land than the small ones because of the economics of its scale.

The relationship between the scale of the farms and the numbers of tractors and other machines is shown in table 3.

The great variation in land ownership (5.5 % of the owners hold about 42.9 % and 94.5 % of the owners hold 57.1 % of the cultivated area) provided that many holders live in towns and agriculture is not their productive activity. At the same time the small holders have an excess of labour

Table 3 Farm Size and Number of Machines in the Egyptian Agricultural Sector

Farm Size	Tractors			Fixed irrigation machines			Portable irrigation machines			Combine-harvesters		
	1	2	3	1	2	3	1	2	3	1	2	3
Under 1 feddan	120	0.9	-	433	3.3	2.1	368	3.1	0.1	425	4.6	0.1
1- Under 2 feddans	144	1.1	-	602	3.7	0.1	318	2.6	0.1	834	9.1	0.2
2- Under 3 feddans	185	1.4	0.1	631	3.8	0.2	329	2.8	0.1	1037	11.3	0.4
3- Under 4 feddans	208	1.6	0.1	742	4.5	0.4	472	3.9	0.3	915	10.0	0.5
4- Under 5 feddans	191	1.5	0.2	626	3.8	0.6	513	4.3	0.5	688	7.5	0.7
5- Under 10 feddans	803	6.3	6.5	2649	16.1	1.5	1943	16.2	1.1	1647	18.0	1.0
6- Under 20 feddans	1707	13.3	2.9	3471	21.0	6.0	2467	20.6	4.2	1057	11.5	1.8
Over 20 feddans	9479	73.8	24.9	7248	43.9	20.3	5570	46.5	15.0	2575	28.1	6.6
Total	12837	100.0	0.7	16507	100.0	1.0	11980	100.0	0.7	9178	100.0	0.5

1) Number of machines
2) Percentage in class to the total number.
3) Number of machine-owners as percentage of total number of holders in each class.

Source: Ministry of Agriculture, General Agricultural Census of 1961, Part I, Section IV, pp. 76-77 (in Arabic).

and other primitive factors of production and want to employ as much as they can. The landlord tenant relations were founded as a result of this shape of land tenure. A great cultivated area was under lease, it was about 43 % of the cultivated area in Egypt in 1972. (2)

The major share of the land leased has been changed from big farms to small ones. This situation means that these farms were distributed among small landless tenants. These farms were not ready to apply agricultural mechanization.

As we have seen above the land tenure in Egypt is the most important factor which affects agricultural mechanization. The gonvernment, the party and the cooperative movement solved the problems in the agricultural sector.

These solutions require modern mechanization of agricultural cooperatives in order to increase agricultural production and productivity.

The new cooperative law made it clear that the agricultural cooperative association is one of our means to protect socialism. (3)

Up to now the shape of land tenure, the classical cooperatives and the insufficient governmental loans of mechanization to the cooperatives are the very important determinants for putting those hopes into practice. Available data on governmental loans for mechanization to the cooperatives amount to 2.4 % of all the loans to the sector and it was not more than 1.9 million LE in 1969. (4)

At present also the personal property holdings are 85.8 % of all the tractors, and the cooperatives hold only 14.2 %. (5)

One of the determinants of agricultural mechanization in Egypt is the limited capital and reserve funds. From table 4 the share of the cooperative in capital was not more than 1477 LE.

Table 4 Agricultural cooperatives: their number and capital from 1952 to 1970

Years	Number of cooperatives	Capital (in thousand Egyptian pounds)	Capital per cooperative (in Egyptian pounds)
1952	1727	661	383
1965	4839	2653	548
1966	4879	2682	550
1967	4921	2801	569
1968	4955	2996	605
1969	4998	3993	799
1970	5013	7404	1477

This amount of capital is too small to cover the various activities of the cooperative and especially in mechanization which needs more capital. The reserve funds are not extensive enough to start mechanization with. In 1970, the share of the cooperative in the reserve funds was only 726 LE. (6) This means that the support of the government in agricultural cooperative mechanization must be one of the tools to encourage it in this field. Governmental encouragement can be given in the long-term and medium-term loans.

We believe that the introduction of agricultural mechanization in Egypt must be related with improving the relations of production which gradually change and improve, without remaining passive in respect to the productive forces, for they open up boundless scope for the development of productive forces.

Large cooperative farms and the cooperative property of the means of production will change this primitive land tenure structure. Without this solution we can expect more small farms, more fragmentation and more uneconomic farms which will tend to decrease productivity and production, further decreases in the annual income of the small farmers, and more inequality of land-ownership and income between the farmers in the country.

References

1 Renne, Roland R., Land Economics. (Harper & Brothers Publishers, New York, 1958), p. 365.

2 Ministry of Agriculture, Department of Agricultural Economics and Statistics: Periodical report 1972, p. 29 (in Arabic).

3 Agricultural Cooperative Law No. 51. of 1969, Chapter I, General rules, Article I.

4 Central Agency for General Mobilization and Statistics, Annual Report on Cooperative Activity in Agriculture 1971 (in Arabic).

5 Central Agency för Accountings.

6 Agency for General Mobilization and Statistics, Annual Report on Cooperative Activity in Agricultural Sector 1973. (in Arabic).

Possibilities for the Improved Production of Meat and Milk in the State Farms of Egypt

by L. Peipp

At the present rate of increase in world population there will be a progressively rising demand for foodstuffs in the decades to come. This is particularly true of the economically less developed countries of Asia, Africa and Latin America. As the social position of broad sectors of the population in these countries improves so will the potential demand further increase and changes in the qualitative composition of the staple food become possible. At the present time it is absolutely necessary to prepare comprehensive measures for surmounting the problem of protein deficiency. There is a particularly lack of protein from animal resources. The present annual population increase of 2.5 to 3 percent. The of general social progress and the ever increasing urban population on the one hand, against a present annual increase in availability of animal products of 1.5 to 2 percent on the other hand result in a rapid widening of the gap in the supply of animal protein. Changes are planned on an international scale as can be seen from the activities on the part of national and international bodies.

The experimental station for agricultural machinery of the VVB Land- und Nahrungsgütertechnik of the GDR, which is jointly managed by the latter and the Egyptian General Organization for Utilization and Development of Reclaimed Land has, after analysing the situation in this area, given particular importance to the possible development of an efficient cattle breeding system. Since August 1973 it has been concerned with studying how to provide safe daily supplies of fresh fodder to large herds of livestock and is also examining ways of achieving storage of fodder by means of suitable preservation methods. With 850 inhabitants per square kilometre of arable land Egypt has an extremely great population density, which is further increasing. Part of the Nasser government plan was the construction of the Assuan high dam. This, the world's largest dam was built and became operative with Soviet aid. It makes a further intensification of agriculture in the Delta possible and at the same time enables the realisation of a large-scale land reclamation programme. From 1975 onwards, 530,000 hectares of former desert and saline soil will also become available. As a rule state farming establishments on reclaimed land cover more than 10,000 hectares of land which is cultivated in centrally controlled state farms using long-term cultivation programmes. An increasing intensification of production is planned for all other state-owned areas. This also includes the large-scale development of meat and milk production areas. In 1973 alone it was planned to increase the number of cattle in the reclaimed land farms from 14,000 to 17,000. The eventual number of cattle there is planned to be 100,000. The programme must make allowance for the specific features of traditional agriculture. In the old valley some 160,000 heads of cattle, 1,250,000 buffaloes and most of the sheep and goats are kept. 95 % of the farms there are smaller than two hectares. Approximately one million farmers in 4,000 villages own 80 % of the dairy cattle. The number of animals per farm varies between one and three. If a cooperative or other form of centralised cattle breeding system is not achieved, the present method of production will continue to prevail. If however a new method were to be achieved then there would be considerable potential for an increase in meat and milk production even in the Nile valley with its dense population of domestic animals. In view of the great importance of this land for food production and for producing urgently needed export crops and raw materials, an extension of the fodder growing area in the intensively farmed region is unacceptable.

Despite good yields from high-yielding varieties Egypt is importing in 1973-1974 US $ 150 million worth of wheat at the present. In 1970 309,350 tons of cotton and 699,696 metric tons of rice were exported.

In the reclaimed land regions of Egypt an efficient animal production system will not greatly affect the amount of land available for the production of food for direct human consumption. With their orientation towards the use of machines, reclaimed land farms are modern agricultural enterprises. Here the idea of large herds presents itself and the only production is that of meat and milk. Moreover, intensive animal husbandry may solve the problems of communities (e.g. the creation of self-contained production units with security of employment). A developed transport system throughout Egypt makes a direct supply to the adjacent overcrowded regions and to the processing centres for animal products possible.

The cultivation of fodder crops in the developed estates is also of basic importance for optimal land utilisation. Clover and other leguminosae are essential to the long-term improvement of the soil. Clover (Egyptian clover, also known as Berseem) is particularly well suited for irrigated cultivation. Each year Egypt grows this crop on nearly 1.2 million hectares. Clower supplies 67 % of the starch and 82 % of the digestible protein for animal feeding. During the period from December to April clover is the sole food. By May 15th, at the latest in all parts of the country all the clover land is ploughed up. This is due to several factors (high temperature, low atmospheric humidity, a lack of water for irrigation, a preference to use the land for summer crops for direct consumption or marketing and phyto-sanitary measures).

The situation as regards animal breeding now changes completely. Fodder has not been preserved. The clover areas are used for grazing. After the fodder harvest in April, the animals begin to suffer from increasing temperatures and under mostly unsatisfactory breeding conditions an acute shortage of fodder arises which lasts until October. The partial overfeeding with clover is abruptly stopped and the feed now consists of straw (barley harvest from 15.4., then later wheat, millet and corn) and concentrates (among other cotton cake). This irregularity can be somewhat compensated in the reclaimed land area by recently increasing cultivation of lucerne. However lucerne is also used for grazing and its growth in the hot season is minimal.

As a result of insufficient animal feeding, the meat and milk production potential throughout the year is not being utilized. On the contrary the animals suffer high weight losses at times. In summer, calf fattening must be severely reduced or completely stopped. The same is true in the case of milk production. In the summer months the dairies receive some 30 % of the milk available at the peak times (the months of February, March, April). This does not even make allowance for the drop in efficiency experienced in animal breeding. Thus, the highest possible use is not made of the animals' potential performance and the periodically available high grade field fodder produced in a region with an extreme land shortage intensive farming.

To overcome the difficulties in the meat and milk production of Egypt it is necessary to establish more efficient herds in the reclaimed land area, which must be kept in large production units under the best conditions. In this connection, completely new requirements will arise for fodder production. It will be necessary to change from extensive pasture feeding to indoor feeding. This will eliminate daily weight losses due to long trails in adverse weather conditions and losses from unbalanced fodder supplies. Fodder crop farming can then profit from the best possible harvest times, optimum frequency of cuts and other factors contributing to an increase in yield and diminishing of losses. Particularly useful organic manure occurs in large quantities in the reclaimed areas and can be spread to suit the demand. At the same time cattle production organized in this way requires the whole year round a supply of fodder balanced both in amount and in quality. Cattle production must be brought into line with crop farming through long-term planning.

Under the given conditions of production in the northern reclaimed land area of Egypt, Egyptian clover will, to a great extent, be the basic fodder. This is in line with a centuries old tradition of Egyptian agriculture. Under favourable conditions it is available as fresh fodder for six months per year. This period up to 4 cuts with yields in each case of 14-24 metric tons per hectare can be obtained. It is possible to keep the animals indoor continually supplied by means of staggering the crops. At the same time, lucerne can also be cultivated in the reclaimed land, being available throughout the year with a total of 6-9 cuts. The yields from green fodder vary per cut between 12-17 tons per hectare. In addition to this the summer supply of fodder should be supplemented by green corn and by several cuts of millet. Any gaps in the supply of fresh fodder must be bridged by means of roughage and silage (Egypt has no experience with silage production as yet).

The realisation of such extensive plans as represented by industry-type enterprises for animal production involves basic changes in management and production methods, compared with the conventional methods. After practical experiences gained over a number of years in Egypt there exist neither insurmountable objective difficulties neither there is a lack of personnel if the many available experts and untrained personnel are trained for their new duties and properly employed.

The experiences so far has been that all fodder plants previously cultivated and most field crops react to a good preparation of the seed bed with uniform germination, better growth and higher yields. Therefore, at the station land for fodder crops was ploughed at a depth of 20-25 cms with the mould-board plough. Tillage with this type of plough uniformly loosens the soil in a single operation. Moreover, the mould-board plough contributes to the organic matter in the soil and in particular helps fight seed-propagated weeds. Organic matter is ploughed in by well-timed ploughing and thus protected from too rapid mineralisation as a result of direct radiation from the sun, as well as from the very high humidity during the night which occurs in the area. The seeds of weeds which have been ploughed under do not break through to the surface. Subsequent final preparation of the seed bed is achieved by using the fine chisel-type cultivator combined with rollers or harrows.

Of particular importance in irrigated land under the conditions of the northern reclaimed areas in Egypt are the careful levelling of the fields as well as a suitable gradient depending on the type of soil and the surface area. Both these factors are essential for an even distribution of water to all the plants in a field. Neglecting this factor results in bare patches in the crop difficulties with the use of machines and finally saline bloom with all its adverse consequences. Efficient ploughing also makes large-scale levelling and clearing of the fields easier. Row seeding is recommended for the fodder crops. In the cases of clover and lucerne the best results are obtained by leaving a space of 10-15 cms between the rows. Row-seeding is also favourable because of the uniform depth of the seed achieved by the machine - an important factor in irrigation agriculture. After the seed had sprouted manure was spread with a machine so as to insure equal distribution. Wheel tractors of up to 90 H.P. are used for the fore-mentioned jobs with the exception of large-scale levveling. The equipment should preferably be mounted because this has the advantage of greater manoeuvrability in irrigation agriculture. For general work the experimental station used the tractors "Nasr 60" (produced in Egypt) and ZT 300 (from the GDR) and the mounted ploughs B 125 with 4x20 Z-bodies and B 126 with 4x30 Z-bodies. Seed drill A 201 (a working span of 3 m) carried out the sowing. This is a universal seed drill which has stood the test in all cases. Clover planted using the above methods on 27.9.73 produced the first green fodder crop on newly cultivated extremely sandy desert soil on 19.12.73. It had been irrigated five times previous to that date (29.9., 5.10., 16.10., 2.11. and 20.11.). For the first cut the average yield (20 random samples) was 15.3 metric tons per hectare. The second cut on 2.2. produced 15.3 tons per hectare. On 19.12.73 plants had grown up to 72 cms and before the second cut to 67 cms. Comparisons between sowing times showed that in the place of later cultivation a drop in yield was experienced. Clover sowed in the same field on 4.10. and 23.10. produced 11.8 and 10.1 tons per hectare respectively. Under the given conditions and in the case of earlier sowing it remains to be seen to what extent increased pest infestation during the early growth stages affects the plants. The severe drop in yield after later sowing could above all have been due to the extreme drop in temperature recorded as early as November and the long cold winter experienced there this year as well as the particularly exposed conditions which were unfavourable to vegetation. The harvesting of the green fodder is carried out by machines. This makes the indoor keeping of animals possible and eliminates energy losses. Moreover, it allows for suppliing fodder to suit the conditions of the animals. It also reduces losses in the fields and avoid reduced regrowth. As regards the daily provision of fresh fodder clover or lucerne may be used long chopped or in its full lengths. From the technical point of view however, a forage chopper was advantages. It cuts the fodder, chops it to the required length and conveys it into a large trailer which is either attached to the machine or else runs alongside it. In our case the forage chopper type E 067 from the GDR was used, pulled and driven by the "Nasr 60" tractor. Trailers from the USSR with 4 tons service load and whose capacity was greatly increased by board attachments were used as transport vehicles. Thanks to the good levelling of the field an uniform cutting height of 6 cms could be used. In the case of Egyptian clover stubble of this height gives the best annual yields. The chopped lengths were at 10 cms. The animals accepted clover and lucerne chopped to these lengths, and in comparison to the longer lengths chowed no great differences in their acceptance of the fodder. Further advantages in the use of the forage chopper are:

(1) Good or very good mowing characteristics, excellent fodder handling and reduced tractor hours and expenditure of labour. In the case of clover with about 19 tons of fresh fodder per hectare, 3 metric tons can be loaded in about 30 minutes (effective working time). Here the labour required is a tractor driver and an unskilled helper. The trailers were attached to MTS-50 tractors imported from USSR.

(2) Optimum payloads if the right attachments are used on the trailers.

(3) Better fodder distribution to the animals. In comparison to long fodder chopped lengths have characteristics similar to those of bulk goods.

(4) Reduction of waste. Immediately after cutting the fodder is conveyed to the transport vehicle in one single operation. Losses occur when there is a strong side wind and fodder can be blown off the trailer. This

is particularly marked when the yields are low as a result of extremely low growth (due to lack of irrigation, among other things).

As yet there have been up to that time no experiences with the operation of self-propelled forage choppers for the daily provision of fresh fodder. On the advantageous side this type of machine would allow greater manoeuvrability in the many small lots of irrigation area, and greater efficiency. In both cases it must be noted, that chopped green fodder is for immediate consumption. It is absolutely necessary to have good coordination of delivery date, demand and extent of supply. If this does not happen then it is impossible to avoid losses as a result of extreme heating of the fodder. Hay making would emerge as a principal mean of fodder preservation in Egypt. Dehydration can be achieved in a variety of ways. The simplest and safest method under the given dry conditions, is drying in the field. With clover hay this can be achieved within 3 days. The disadvantage in this, however, is the extraordinarily high loss of dry leaves. As a result of low atmospheric humidity and the hot sun the leaves dry notably quicker than the stalks. Thus it is not uncommon for the stalk to be not yet dry while the leaf falls off at the slightest movement of the swath (e.g. while being gathered). In the case of clover and lucerne this means the unavoidable loss of valuable fodder, but this can at least be reduced by careful handling of the dry material. For example this is made possible by handling the crop in the early morning and by using stalk breakers. The latter damage the stalk tissue and thus helps the water to evaporate so that the difference in drying time between the stalk and the leaf can be reduced.

Efficient hay making is only possible by using machines. In field drying good results can be achieved with the mower E 301. This is a self-propelled machine with a cutting span of 4.20 m. It cuts the fodder, passes it through stalk breakers, and lays it in swaths. Fodder cut using the mower E 301 frequently requires no further treatment because the swath is very loose. There is sufficient access for air to all parts of the plants. Once the storage humidity is reached than the hay is baled by using the bailer K 442. Only hay that is moist with dew can be baled with little loss and is storable. Other methods of achieving dry crops are the harvesting of wilted fodder with the subsequent airing in a store and the hot air drying of the chopped fodder. Both methods result in minimal losses and a far higher quality fodder than achieved by field drying. At the present time more exact research into the air drying of prewilted clover and lucerne is being carried out. Today it is safe to say that soon by making use of available possibilities including the use of more efficient fodder harvesting machines break down in animal performance due to lack of fodder can be avoided. From the point of view of fodder production, conditions can be created in Egypt in which more animal produce becomes available.

Pedological Aspects of Mechanized Cultivation of the Soil in the Tropics

by H. Pagel

1. Introduction

Measures for the intensification of agriculture are aimed essentially at making more effective use of the soil, the basic mean of agricultural production. More effective use of this means of production must, however, include the conservation or improvement of soil fertility in order to maintain and increase yields.

Mechanization and agrochemicals are important factors of the scientific and technological revolution in agriculture. However, like all techniques related to soil utilization, they are not restricted to scientific and technological aspects but have important economic and social implications, which are dependent on the state of development of the productive forces and on the social system.

The mechanization of soil cultivation is an integral part of the greater complex of farm mechanization. Under certain economic and social conditions the mechanization of soil cultivation, in comparison with other measures of mechanization (e.g. harvesting) is of special significance, since it frequently represents on the one hand, a sort of initial step of further mechanization (similar to the use of mineral fertilisers) and on the other, not only increases yields, but often becomes a prerequisite for extending the acreage under crop.

Social development, the intensification of agriculture and the use of scientific findings require the mechanization of soil cultivation in the tropical and subtropical countries. As to the "how" with regard to mechanized soil cultivation in the tropics, there is certainly still an abundance of questions to be asked (Franke, 1968; Voss, 1974); the scientific approach to the problems of cultivation in the tropics has suffered not only from the small number of cultivation trials but also from the insufficent description and classification of the soils on which these trials have been carried out. As a result of this, and the greater variability of soils in tropical climates and the more precarious equilibrium there between climate, soil and vegetation, there are still many questions about soil cultivation in the tropics that remain unanswered. Our knowledge about conditions in the temperate climates is not really transferable to the tropics and accordingly, not only successes and increases in yields are reported, with more intensive and more mechanized soil cultivation, but also damage such as structural deterioration, erosion, lower yields, devastation etc.

From the manifold problems, three will be dealt with from the viewpoint of soil science

1. Soil structure as a prime factor in the cultivation of soils;
2. Variations of soil structure resulting from tillage;
3. Advantages of mechanized cultivation in the tropics.

The primary target is to relate the important soil qualities, which are to be taken into consideration during cultivation, to the basic genetic units and thus to give those in charge of mechanization or the technologists data for using the available soil maps, with regard to the necessity and to the problems of cultivation of soil groups.

2. Soil structure as a prime factor in soil cultivation

Any method of soil cultivation influences and alters primarily the soil structure, that is to say, the spatial arrangement of the solids soil constituents. The water and air regime and soil temperature depend directly on the structure. These influence on the other hand, many chemical and biological soil properties. Soil structure is accordingly a vital factor in effective soil cultivation.

The basic elements of structure are the a g g r e g a t e s, which are formed through coagulation and bonding and are destroyed by aggregate destruction and peptization. Cultivation can lead on the one hand (in compacted soils) to bonding, and on the other, can mechanically destroy the aggregates which are present. This is the primary point of soil cultivation. The natural stability of the aggregates and their resistance against destruction, depend on the bonding substances. The most important are:

1. Oxides and hydroxides of Al, Fe, and Mn. They are typical bonding substances in the red and brown soils of the humid tropics (in ferrallitic soils); the structure of these soils is characterized by high porosity and great stability. Their cultivation is possible within a wide range of moisture contents (contrary to minute soils) and their breaking up requirement is very low, as a result of the high porosity (stabilized structure). These soils are common in the humid forest regions e.g. in Ghana, Tanzania, Cuba, Vietnam, India, Bangla-Desh, Burma, under conditions of free drainage.

2. As a further bonding medium clay minerals are to be found. This is particularly the case in vertisols but also in all clay soils with single grain and coherent structure (e.g. alluvial or gley or alkali soils). The structure of vertisols is characterized by relatively low porosity (high density) and by marked structure variations depending on the moisture content (stone - hard if dry and muddy, plastic if wet). Alluvial and gely soils are common in river plains, vertisols in plains (Sudan, India, Iraq, Syria, Bangla Desh, Mali etc.), alkali soils are common especially in semi-arid regions, e.g. Iraq, Syria, Egypt.

3. In all soils organic matter (humus) is to be found as a bonding substance. Porosity then depends on the humus content; the structure is more stable than with clay mineral bonding, but less stable than with oxide bonding. Especially in fersiallitic soils and in sub-arid soils humus content, is a decisive factor. Such soils occur especially in the alternating humid tropics and subtropics (Tanzania, Somalia, Ghana, India) and in the semi-arid regions (e.g. Iraq, Syria, Somalia).

The necessity and effect of cultivation therefore clearly depend on the prevailing bonding substance. They depend even more on the structural form of the soil. For ferrallitic soils, the crumb structure is typical in vertisols, the polyhedron and prismatic structure prevails; this applies to alkali soils as well. In the A - horizon of desert soils and other alkali soils, the single grain structure is, on the other hand, frequent; a platy structure is frequently found in soil compacted by cultivation (plough pan).

From the prevailing bonding substance and the structural form results structural stability. Experiments carried out by Meymard and Combeau (1960) indicated the following sequence of decreasing structural stability for some tropical and subtropical soils (Fig. 1): ferrallitic soils > fersiallitic soils > vertisols > alluvial soils and gleysoils > alkali soils.

Allowing for other soil properties, the following may be deduced by simplification:

1. The necessity of loosening the soil is slight in ferrallitic soils, e.g. in Ghana, Cuba, but increases toward the alkali soils (e.g. in Iraq, Egypt, Sudan).
2. The necessity of intensive cultivation (ploughing) is great in alkali soils and minimal in ferrallitic ones.

This general pattern can, of course, be broken by particular structural conditions or particular requirements of plants which are grown; e.g. in fersiallitic soils, with a compacted B - horizon (as a result of leaching), deep loosening may be appropriate, and for the cultivation of rice in alluvial soils mechanical compaction may be needed.

3. Variations of soil structure resulting from tillage

The sum effect of tilling is normally a loosening or compaction of certain soil layers, mostly the surface ones, i.e. an increase or decrease in porosity. Not all pores, however, are affected to the same degree:

1. The "fine" pores which contain non-available water are practically not affected.
2. The "medium" pores, which contain water or air available for plants may be somewhat increased or decreased.
3. The most marked variation occurs with "coarse" pores, which permit ventilation, permeability, and water seepage.

Soil loosening, therefore, means a rise in the proportion of coarse pores and soil compaction nearly always reduces the proportion of this pore size.

The dependence of the medium pore size distribution on the texture shows that sandy soils will require little loosening e.g. in the dry regions of Iraq, Syria, Egypt, Mali, and that clay soils (e.g. Sudan, Egypt), will have a high soil loosening requirement (often identical to tilling requirement). This general statement, however, becomes invalid in the humid tropics because of the process of "Vererdung", or "pseudo-sand formation" in ferrallitic soils. Here the individual clay particles bake together into greater aggregates as a result of oxide bonding and the soil becomes "sandy" despite the extremely high clay content, which is anything up to 90 %, e.g. in Cuba, Vietnam, Ghana. Unfortunately, this is not true for all ferrallitic soils. The significance of texture for the pore size distribution and cultivation is therefore strongly masked by the character of the genetic soil group.

In the ferrallitic soils of the tropics which have a high natural porosity, repeated loosening (ploughing) will rather cause compaction and a reduction of porosity; this is a self-defeating approach. Deep ploughing, for the purpose of soil loosening, is pointless, promotes the decomposition of humus and reduces structural stability.

The changes in structure represented in Figure 2, of a ferrallitic soil in Central africa under shifting cultivation (according to Combeau and Quantin, 1963) clearly shows, that:

1. the structural stability is markedly reduced with continual cultivation or during the period of cultivation,
2. structural stability and the related porosity again increase during the fallow period,
3. the rate of structural deterioration is greater than structural re-improvement.

Tilling is not the sole factor responsible for this. The removial of the vegetation and the natural soil cover, humus breakdown and the direct influence of atmospherilia play a certain role. The consequences of intensive cultivation on these soils are often marked erosion and a decrease in fertility. The deterioration of the structure of virgin soils with increasing cultivation was related by Bradford and Miller (1954) to three causes:

1. The destruction of the original vegetation cover and the direct effect of the weather on the soil surface resulting from it (drying up, mud, compaction, erosion).
2. The mechanical destruction of the aggregates by tilling implements.
3. The loosening (increase of the proportion of coarse pores), and ventilation promote the oxidation or the decomposition of organic matter, with the natural polymeric organic compounds which stabilize the aggregates subject to a greater degree of breakup than organic matter as a whole.

Accordingly, soils with prevailing organic bonding substances should show more structural deterioration than ferrallitic soils. Boyer and Combeau (1960) found a close relationship between structural stability and the content of organic substance in fersiallitic and ferrallitic soils without pseudosand in Central Africa. Even after short periods of use and cultivation (a single ploughing with subsequent grubbing and harrowing) there occured a marked structural deterioration (compaction and muddiness).

The preservation, as far as possible, of the vegetation cover, the preservation of humus by crop rotation, ground cover or mulching and minimum tilling, are certainly the most important measures for preserving a favourable structure and a high structural stability in the red and brown soils of the humid tropics; a good ground cover can further reduce the cultivation requirement of these soils, which is already low.

Even in soils with compaction phenomena (leached fersiallitic soils, alkali soils, vertisols, alluvial soils), mechanical loosening should be combined with biological measures (plant cultivation, cover).

In arid regions which are not irrigated, (sub-arid soils, partly semi-desert soils) the replacement of the extensive digging kind of soil cultivation by intensive ploughing, after an initial increase in yields (loosening, water storage), leads frequently to more marked wind and water erosion, to structural decay and "devastation" (sandiness), since the content of organic matter as the most important bonding substance, is low and is even further reduced with intensive cultivation. Stubble mulching, minimal tillage and digging instead of upturning would seem appropriate here.

In highly compacted soils (e.g. many soils of an alkali nature in India, Iraq, Egypt), in soils with an unstable structure, (many soils after a long period of cultivation) in clayey alluvial and gleysoils (e.g. in Iraq, Sudan, Egypt, Bangla Desh), intensive soil cultivation is, on the contrary, a prerequisite for higher yields; while it does not increase structural stability of the soil, it certainly improves the degree of looseness.

4. Advantages of mechanized soil cultivation in the tropics

1. The possibility to cultivate a higher number of acres per unit of time enables tillage to be done at the right time; from the pedological standpoint, this means that the soil can be cultivated at a definite point in time when it has the optimal water content. This is especially significant for vertisols, e.g. in Sudan, which in comparison with the clayey "minute" soils of the temperate climates can be described as "second" soils.

But even in the other soils, tilling at the right time can lead to substantial increases in yields. Voss (1974) mentions examples from Kenya, Zambia, Turkey and other countries.

2. Greater (mechanized) traction power enables the combination of implements and a reduction in the number of tilling operations. This can promote the quick growth of a plant cover and therefore reduce the "barren" period; the possibility of planting several crops during the year becomes greater.

3. Certain soils (especially vertisols, partly the so-called planosols) require motor traction and strong tilling implements for crop farming on a larger scale; in this case mechanized cultivation is a prerequisite for extending the area under crop (e.g. in Central Sudan, or in Ghana south of the Volta Dam).

4. The deep loosening (of soils with compaction in the profile), and the systematic deepening of the surface soil (in strongly leached soils, in podzols and solod-soils) are possible only by mechanized soil cultivation.

5. The improvement of some soils (e.g. the mixing of the calcareaus B - horizon with the lower part of the A - horizon in alkali soils, the deep loosening of saline low-land soils for the purpose of improved salt washing,

e.g. Egypt) is only possible to a greater degree by mechanized soil cultivation.

The advantages of mechanized soil cultivation, with the further development of tropical agriculture, are thus clear from a pedological standpoint.

The stronger effects of mechanized soil cultivation (mechanical effect) on the structural aggregates together with a lack of scientific understanding and practical experience increase the possibility of errors in practical tilling operations. The consequences then are the destruction of favourable structural conditions, a decrease in structural stability, an increase in water and wind erosion and finally a decrease in the fertility and in yields. In order to avoid frequently occurring mistakes, the following point should be considered:

1. Mechanized soil cultivation requires the simultaneous use of other techniques of mechanization and other measures for increasing yields (fertilization, use of herbicides, protection against erosion, crop rotation etc.).

2. Mechanized soil cultivation has partly led to the tilling and agricultural utilization of critical areas in the humid tropics (extreme ferrallitic soils e.g. in Cuba, Brazil) and arid tropics (semi-arid regions in Central Asia). In these cases it should be decided, after examining all the factors involved, whether the cultivation of aborescent and shrublike perennial crops (on ferrallitic soils in the humid tropics) or the utilization as pasture land (on sub-arid soils in the arid tropics and sub-tropics) is possible and economical. If cultivation is considered, then the number of tillige operations should be reduced to a minimum and soil cultivation should be combined in ferrallitic soils with ground - covering (mulching) and in arid regions with stubble mulching and other methods of "dry farming". In arid regions a digging kind of soil cultivation can be superior to upturning.

3. The necessary working of greater areas and its possible effects on water and wind erosion can be offset by suitable crop rotation, by cultivation across the slope, by a ground cover, by strip cultivation and shelter belt strips (in arid regions).

4. In regions with ferrallitic soils and in arid regions with rain-fed cultivation, the primary consideration in mechanized tilling should not be the substitution of the local and traditional implements by the upturning plough but mechanized traction and such tilling implements which have effects roughly similar to the traditional implements (e.g. the grubber, disc harrow and others).

5. In many cases the need for tilling operations can be minimized by the combination of soil cultivation and chemical weed control.

For the full utilization of the advantages of mechanized soil cultivation, more experiments and observations are necessary in the humid and arid tropics on a large scale.

5. Conclusions

1. Experiences from the temperate climates with regard to soil cultivation are not readily applicable to tropical soils.

2. The type of bonding matter for aggregate formation, the structural form, the pore size distribution and structural stability are essential factors affecting the necessity and effect of soil cultivation and should therefore be taken into consideration.

3. With a degree of simplification these factors can be characterized for certain genetic soil units or assigned to them.

4. Using this assignment people in charge of mechanization or technologists can derive general information regarding appropriate soil cultivation from available soil maps showing the extent of the area of these genetic soil units.

5. The advantages of mechanized soil cultivation in the tropics can only be fully utilized, and mistakes avoided, if tilling is carried out in conjunction with other measures of intensification with due regard for ecological (soil), economic and social conditions.

Literature

Boyer, J. and A. Combeau (1960), A study of the structural stability of some ferrallitic soils from the Central African Republic. Afric. Soils, Vol. No. 1. 25 - 42.

Bradfield, R. and R. D. Miller (1954), Soil structure. Tranc. 5 th Int. Contr. Soil Sci., Leopoldville 1954, Vol. 1, 131 - 145.

Combeau, A. et P., Quantin, Observations sur les variations dans le temps de la stabilité structurale des sols en region tropical. Dah. Pedol. ORSTOM 3., 1963, 17 - 26.

Franke, G. (1968), Lehrbriefe Tropische und Subtropische Landwirtschaft. Ackerbau, Heft 1, Leipzig 1968, S. 99 ff.

Maymard, J. and A. Combeau (1960), The residual effect of submersion on soil structure (demonstrated by applying S. Henin's index of structural stability). Afric. Soils, Vol. No. 2, 141 - 148.

Voss, C. (1974), Agricultural mechanization, production and employment. FAO Monthly Bulletin of Agricultural Economics and Statistics. Vol. 23, p. 1 - 7.

Change of the structure of a ferrallitic soil under shifting cultivation in Central Africa

(after COMBEAU and QUANTIN, 1963)

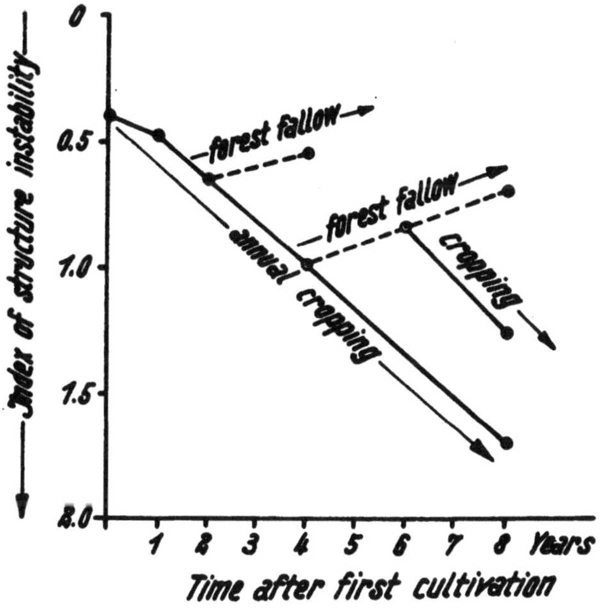

Stability of the structure and permeability of some tropical soils

(after MAYMARD and COMBEAU, 1960)

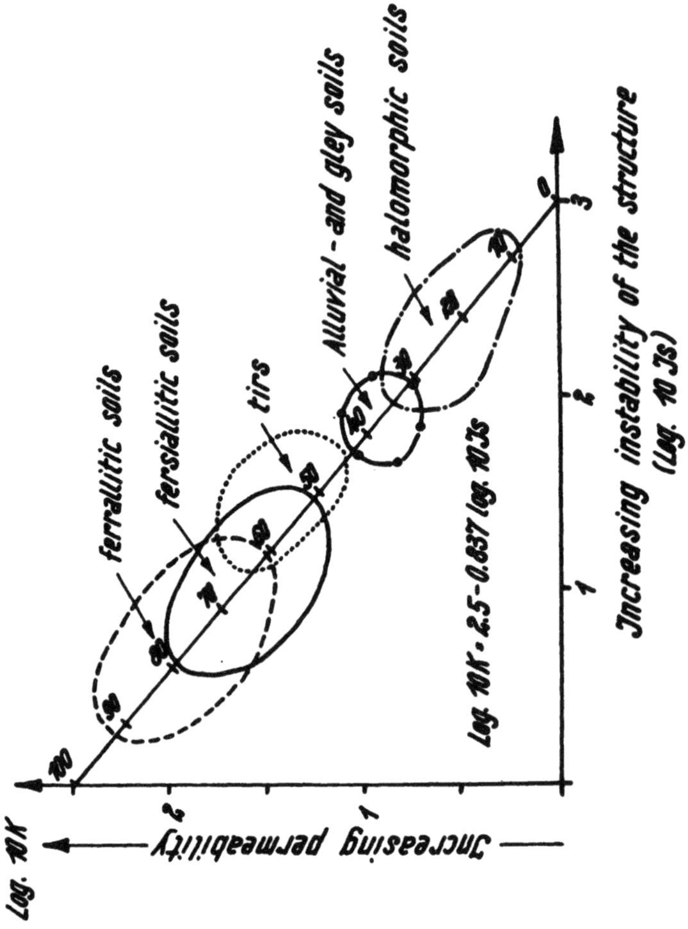

Mechanization of Plant Protection in Developing Countries

by G. Fröhlich

The importance of mechanization of agricultural production in developing countries is among the principal topics of discussion at this scientific conference. Mechanization of crop protection should be considered an integral part of all processes of mechanization of land reclamation, cultivation of soil, irrigation, soil conservation, harvesting, and processing. The use of modern means of agricultural production including high-yielding seeds and seedlings, fertilizers, tillage equipment, cultivators, and harvesting machinery, which provide for both healthy plant growth and lossless harvesting, are not usually sufficient to collect and carry in potential crop yields, as these may be decimated by various environmental influences including parasites causing plant diseases, plant pests, and weeds.

The amount of losses suffered may be different in various regions as well as for the simple plant species. Rough calculations by Cramer (1967) show that the losses caused by biofactors in wheat production in the world reach about 24 % of the potential yield, which will come to approximately 85.6 million tons of wheat, an amount which would be sufficient for roughly 45 million people to be supplied with bread grain for one year. In rice production, the losses suffered, for example, by India and Indonesia are in the neighbourhood of 57 %, whereas rice losses suffered by Japanese growers are only about 14 %. In cotton production, losses ranging from 30 to 40 % of potential yields have been reported. Similar estimates could be made for many other species of plants and regions, but we may refer to the publication by Cramer (1967), which gives a wealth of relevant data.

The losses given above and in the paper by Cramer (1967) are based upon estimates made with the present state of agricultural production taken as a base. However, it has been found in practice that an intensification in agricultural production is not only accompanied by an increase of yields, but also by an increasing danger of the plants being affected by pests and diseases. Detailed studies of this problem were made by workers of WACRI at Tafo, Ghana,in respect of cocoa production, to which particular reference sould be made here. However, it would be unreasonable to assume that the rate of infestation with plant pests and diseases increases in a parallel relationship with intensification of agricultural production. Rather, it is usually possible for a certain constancy to be observed after an occasionally very rapid increase in rate of infestation, which, for a number of plant pests, may be regarded as being connected with their radii of action during infestation. Such plant pests, aphids, gall midges, scale insects, white flies, weevils, etc. - invade fields and plantations from outside and, because of their limited flying activity, usually settle themselves only in 25 to 50 m large marginal zones. If the fields or plantations are of a correspondingly large size, then this will necessarily reduce the probability of infestation. For example, if, in the case of infestation of a 25 m marginal zone of a 1 hectare large field having the form of a square, overall infestation is 75 %, then the latter will be only about 10 % for a 100 ha field having the same quadratic form (Wellmann, 1937; Kühne, 1967). However, the danger of sizable losses exists here, too, so that it is imperative that plant protection measures should be taken with a view to protecting cultivated plants from infestation with plant pests and, thus, contributing to the numerous efforts made to insure potential crop yields.

The possible crop protection measures are multifarious. Unfortunately, however, full use is not always made of the various methods of plant protection that are available. The mechanization of crop protection requires - aside from some physical methods,such as trap ditches, trap slots, and similar means - above all the use of equipment and machinery that may be employed for the application to plants of both,chemical pest control products and biological pest control agents (e.g. entomobacteria). There are many types of equipment and machines that may be used for this purpose, and they range from manually operated equipment through motor-driven machines to airplanes which are either fixed-wing planes or helicopters. The concern of this paper is, not to discuss the advantages and disadvantages of the various forms and types of plant protection equipment presently available, but, rather, to deal with their economic use under the special conditions of agricultural production in developing countries and also consider the measures required as a prior condition to such use.

During our practical activities in the tropics and, more particularly, in regions where agricultural production had already risen above the level of mere subsistence farming it actually was never necessary for us to discuss the need for, and importance of, plant protection, which had already been propagandized by the plant pests themselves. The possibilities of control pests and diseases were always a prime consideration in our discussions with farmers. Although they always notified their willingness to cooperate effectively in agricultural production we very frequently met with an individualistic mode of thought, in that smallholders often tried to deal with the problem of pest control by their own even they could get knapsack sprayers on loan from the cooperative. Unfortunately, however, pests and diseases do not follow this way of thinking. For them there are no such things as individual fields or field borders, and when they epidemically occur in large masses, they usually invade, within a relatively short time, all fields and plantations planted with one or several types of plant in one area. They may be effectively controlled only if the defense measures taken are applied within a short time and extended over the entire area invaded by them. This shows quite clearly that an effective control of pests and parasites cannot be carried out by the individual farmer alone, which, in some places, is still done today especially under capitalist conditions, but should be carried out by all farmers of the area within the framework of a concerted action. The consequences of a policy leaving pest control to the individual farmer may be exemplified by reference to the invasion of Spanish citrus fruit growing areas by the Mediterranean fruitfly (Berville, 1969; Koppelberg and Cramer, 1969). Here, export losses of 2 to 2.5 % of harvested crops,

which, moneywise, amount to some 5 to 10 million marks, were suffered year in and year out until measures taken by the government and providing for tightly organized and centralized pest control by a combined insecticide and bait applied to pest-infested areas with the use of airplanes enabled previous drastic losses to be reduced to 0.2 to 0.5 % or, moneywise, 1 to 2.5 million marks. From the example shown above it is evident that for the control of pests and parasites occurring in large masses to be successful it is absolutely necessary that all of the control measures taken be extended over large areas and be applied with sudden success. What do we understand by control measures applied with sudden success and extended over a large area? This requires treatment of all pest-infested fields of a region within a very limited period of time and with the aid of agents and equipment producing a high biological effect and destroying, in a short time, 80 to 90 % of the pests and parasites. This desired biological effect is largely dependent upon the degree of distribution of pesticides within the field or plantation as well as over individual plants. However, it is not usually possible for a high degree of distribution to be obtained by means of manually operated equipment, it rather requires relatively high pressures which may only be obtained from motors or, else, through power takeoff shafts. Added to this comes that previously widely used spraying methods requiring the use of large amounts of water and employed, for example, for controlling the Sigatoka disease in banana plantations the amounts of water required for this purpose being in the order of 3,000 l/ha, are, for economic reasons, gradually being replaced by water-saving methods. The most important methods of application in field production and fruit growing include:
High- or medium-volume spraying - 200 to 600 l/ha
Low-volume spraying including drift spraying - 50 to 400 l/ha
Ultra-low-volume and ultra-ultra-low-volume spraying - 3 to 20 l/ha
Mist blowing or fogging - 3 to 6 l/ha
Dusting - 8 to 20 kg/ha
Application of granular substances - 5 to 10 kg/ha.

The number of reports covering the successful use of ultra-low-volume spraying have greatly increased during the past few years. Mention should be made in this particular context of a number of papers published by Matthews (1973) reporting on the control in the Republic of Malawi of cotton pests, Tanaka (1972) describing the control in Japan, of rice pests, Bindra, Singh & Chaudhary (1970) presenting the results of control, in India sugar-cane of pests and Wolfenberger & McGarr (1971) reporting the successful control of cotton pests.

All of the above papers emphasize the favourable distribution of active substance over the plants in addition to the considerable saving in water by this method. However, the smaller the amount of water, the higher the percentage of active substance in the emulsion or suspension being applied since it is the amount of solvent rather than the amount of active substance per hectare that is drastically reduced in ultra-low-volume spraying. If it should not be possible for an extremely fine distribution of biocides over the plant to be obtained, perhaps by using an inadequate technique of applic-ation, then phytotoxic damage will be produced which, in the end, will lead to losses similar to those that would have been caused by the pests destroy-ed. Consequently, the success of spraying is governed by the use of suitable plant protection equipment and machinery fitted with appropriate ultra-low-volume spraying units. They include knapsack-type motorized ultra-low-volume sprayers, tractor-mounted or trailed sprayers, as well as air-planes that are equipped with suitable spraying nozzles. Selection of the equipment or machine to be used is governed by the size of area to be treated. A paper by El-Tegani (1974), to be presented at this seminar, shows that, in the Republic of the Sudan, about 400, 000 hectares of cotton are treated six times a year with plant protection products destroying the most important cotton pests. To 90, 000 hectares of land under control products were applied by airplanes using the ultra-low-volume method. In addition, tractor-mounted spraying equipment is also being widely used, whereas the use of knapsack-type motorized ultra-low-volume sprayers and Landrover-mounted drift sprayers is generally limited to very small areas. On the other hand, the Gezira Scheme, by large-scale coordination of cotton cultivation, provided for the creation of crop acreages permitting airplanes and larger-size machines to be used economically. Unfortunately, however, such favourable conditions may be found only in a very small number of agricultural regions in developing countries. Of course, the use of large-capacity machines giving the desired performance requires that larger cultivated areas be available. For example, a tractor-mounted sprayer developed in the German Democratic Republic and having a capacity of 600 liters requires, for a high annual rate of utilization, an area of some 900 to 1,000 hectares, whereas a trailed sprayer developed in close cooperation with Hungary and having a capacity of 1,000 liters requires an area of about 2,250 to 2,500 hectares.

It is quite obvious, therefore, that effective crop protection requires not only the availability of suitable chemicals and corresponding equipment for their application, but also an adaptation thereto of crop farming methods, in order that the former may be used profitably. The Gezira Scheme example has already been referred to. The method of cooperative crop rotation developed in the Arab Republic of Egypt is another example of how such preconditions may be created even if the level of development of cooperatives is still rather low. The same is true of the forms of cooperative land use in Tanzania's Ujamaa villages and numerous similar initiatives taken in other developing countries, which should be regarded as initial efforts made along this line.

Of course, plant protection is also possible with the use of knapsack-type motorized sprayers, this practice of spraying having been observed by us in Ghana, Mali, and other countries; as far as perennials are concerned, the present stage of development often allows no other method to be used, which is due simply to the fact that changes in coffee, cocoa, or banana production usually require extensive preparations involving a relatively long period of time. For all annuals, however, the use of such equipment appears to be less suitable.

This may be illustrated by the following example:

The farmers of an Iraqi cooperative usually have more than 2,000 hectares of land available to them (Michalski, 1973). If we assume that about 500 hectares are used for cotton cultivation, i.e., for growing a crop requiring intensive protection, then, by using the formula set up by Dünnebeil (1974) for the planning for plant protection equipment to be used on large areas under crop and with sudden success the following number of machines would be required:

$$M = \frac{F}{Z_1 \cdot Z \cdot t}$$

where F = crops to be treated, hectares

Z = hours of machine use per day

Z_1 = length of time for biologically effective control, this being the period of time during which a plant pest may be most effectively controlled

t = average area covered in unit time, ha/h

Let the hours of machine use per day be 8 and the length of time required for biologically effective control be about 3 days in this example.

The types of equipment to be compared include the following:

1 knapsack-type ultra-low-volume sprayer with an area capacity of 0.3 ha/h;

1 tractor-mounted sprayer equipped with a 600-l container and having an area capacity of 2.4 ha/h; and

1 trailed sprayer equipped with a 1,000-l container and having an area capacity of 6.5 ha/h.

The number of units required for 500 hectares of land under cotton cultivation would be approximately 70 knapsack-type sprayers, or about 9 tractor-mounted sprayers, or circ. 3 trailed sprayers.

Six annual treatments of the cotton fields would result in only about one third of the capacities of the tractor-mounted and trailed sprayers (the annual capacities of which are of the order of 350 and 1,000 hectares, respectively) being utilized in cotton cultivation, so that it would be possible for them to be used also for the protection of other crops (weed control in grain fields, pest control in fruit plantations). Since it is often assumed that the capital cost of smaller-size equipment is lower than that of larger types of equipment, the comparison which is presented here is made on the assumption that effective measures of defense can and will be taken on a large scale:

Cost of 70 knapsack-type sprayers: c. 84,000 marks
Cost of 9 tractor-mounted sprayers: c. 81,000 marks
Cost of 3 trailed sprayers: c. 75,000 marks

It is evident from this comparison that capital costs do not increase with increasing size of plant protection equipment. The problem rather lying in a rational utilization of available equipment, which in general requires that corresponding acreages should be available. Added to this comes that about seventy qualified operators are required for the efficient use of the above-given number of knapsack-type sprayers, whereas the number of skilled operators is about 18 for the 9 tractor-mounted sprayers and as small as 6 for the 3 trailed sprayers. It should be emphasized again that only highly skilled labour should be employed for the carrying out of all work to be done in connection with crop protection, since the use of poisons and the handling of the above-described machinery and equipment require that pest control operators be fully familiar with the relevant occupational safety and health standards (including the tolerances in regard of residues in crops grown for food). From the examples shown above the following general conclusions may be drawn:

(1) Intensification in agricultural production, especially in the case of cultivation of exceedingly delicate high yielding varieties, usually result in increases in plant pest and parasite populations, the control of which should be properly planned and organized in good time.

(2) The successful control of plant pests and parasites is only possible within the framework of concerted actions, since separate actions will not give the desired result.

(3) Control measures should be taken cooperatively, immediately, and extended over large areas in order for them to be successful. To this end, an adequate number of plant protection equipments are required.

(4) The biological effect to be produced for the successful control of pests requires that the pesticides be extremely finely distributed over the entire crop area as well as the individual plants, which cannot usually be achieved with the use of manually operated equipment.

(5) The purchase of motor- or shaft-driven plant protection equipment including airplanes calls for capital projects being financed either cooperatively or by the government. Such equipment may be profitably utilized only if there are correspondingly large crop areas realized by cooperative cultivation of certain types of plants, cooperative crop rotation, block farming, or similar forms of crop farming.

(6) Crop protection measures cannot be left to individual farmers, since all plant protection requires that well-qualified pest control operators be employed. The organizations charged to carry out pest control should be:

- technically well-equipped teams of pest control operators such as those organized, for example, in the Republic of the Ivory Coast (Gry, 1954), or
- special teams of pest control operators such as those organized within the framework of the Gezira Scheme, or
- teams of pest control operators recruited from and working for farmers' cooperatives, or
- plant protection organizations constituted by different cooperatives working together that could gradually develop into agrochemical centres.

The duties of such teams would include the supervision and inspection of areas under crops, the planning for plant protection products and equipment, the maintenance of equipment and machinery, and the use of plant protection products and equipment whenever pest control measures have to be taken. Although privately operated crop protection organizations, a number of which are run in developing countries, have occasionally been rendering valuable services, they have a major disadvantage in that they are chiefly concerned in making profits and generally pay only little attention to the economic interests of their partners, the individual farmers or farm cooperatives.

Where such private operators are allowed to contract to do plant protection work, the government should be in control of their activities. What we recommend is the establishment of plant protection organizations operated by either the state or farmers' cooperatives.

Literature

Berville, P. (1969), La lutte contre la mouche des fruits en Espagne. Phytoma 21, 35-36.

Bindra, O.S.; H. Singh & J.P. Chaudhary (1970), Ground application of ultra low volume sprays for the integrated control of sugarcane leaf-hopper, Pyrilla perpusilla Walker. Ind. J. Ent. 32, 4-6.

Cramer, H.H. (1967), Pflanzenschutz und Welternte. Plant protection and World Yield. Bayer "Pflanzenschutz", Leverkusen.

Dünnebeil, H. (1974), Pflanzenschutztechnik und Einsatztechnologie. Technology of plant production and application, in: Fröhlich et al. (1974) "Pflanzenschutz in den Tropen."

El Tegani, M. (1974), Report on conventional and ultra-low-volume spraying and formulations. Agric. Res. Crop Wadi Medani.

Gry, J. (1954), La lutte contre le parasitisme des cultures en Côte d'Ivoire.

Koppelberg, B. & H.H. Cramer (1969), Die Bekämpfung der Mittelmeerfruchtfliege, Ceratitis capitata Wied., in Spanien, Verfahren und wirtschaftliche Bedeutung. Control of the Mediterranean fruitfly, Ceratitis capitata Wied. in Spain, procedures, and economical importance. Pfl.schutzn. "Bayer" Leverkusen 22, 164-174.

Kühne, W. (1967), Zur Befallsverteilung der Kohlschotengallmücke (Dasyneura brassicae Winn.) in großflächigen Beständen. Beitr. Ent. 17, 287-297. Distribution of infestation by Dasyneura brassicae Winn. on large areas.

Matthews, G.A. (1973), Ultra-low-volume spray application on cotton in Malawi. PANS 19, 48-53.

Michalski, K.-J. (1973), Landwirtschaftliche Genossenschaften in afro-asiatischen Entwicklungsländern. Akademie-Verlag, Berlin. Agricultural cooperatives in Afro-Asian developing countries.

Tanaka, T. (1972), Experiments on ULV ground application in paddy fields. Jap. Pest Inf. 13, 11-15.

Wellmann, T.L. (1937), Control of southern celery mosaic in Florida by removing weeds that serve as sources of mosaic infection. U.S. Dept. Agr. Techn. Bull. 548, 1-16.

Wolfenbarger, D.A. & R.L. McGarr (1971), Low-volume and ultra-low-volume sprays of malathion and methyl parathion for culture of three lepidopterous cotton pests. Prod. Res. Rep. USDA 126, 1-14.

Suitability of Farm Tractors and Farm Equipment for Tropical and Subtropical Environments — Problems of Climatetechnology

by J. Jára

More than three fifths of the potentially productive arable land of the world are in the tropical and subtropical regions, and most of it in the developing countries of Asia, Africa, and South America. On the other hand, farm tractors and equipment have been manufactured for the most part outside the "Third World", and imported here on a large scale. Efficient operation and performance levels attained by imported farm machinery under tropical and subtropical conditions would be governed, when viewed from the viewpoint of technology, to a considerable extent by its suitability to operate under different climatic, production, and economic conditions.

The science of climatotechnology (termed formerly tropicalization) is concerned with a number of factors affecting output, efficiency, service life, and suitability of farm tractors and other farm equipment and implements imported to countries featuring tropical and subtropical climates. It should be borne in mind, however, that the number of the factors involved is greater than that relating to other mechanical engineering enterprises. The factors in question may be grouped as follows:

(1) Climatic factors,

(2) Production and crop management factors,

(3) Effect of operation,

(4) Effect of transport and storage.

1. Spoilage of Farm Machinery due to Climatic Factors

Contrary to other mechanical equipment, farm tractors and farm machinery would be exposed generally throughout its operating life to extreme climatic conditions, suffering both from sunshine, rain, high temperature, temperatur variations, abrasive and corrosive effects of the soil worked, high dust and sand percentages carried by the air currents, high air humidity, etc. The factors listed above not only affect the equipment, i.e. its material and life, but also its operating conditions and parameters.

Some typical examples of machine and machine parts degradation caused by climatic factors are given below:

a) Atmospheric corrosion of metals and metallic parts - The mechanism of the corrosion phenomena has been theoretically well understood, including the factors that do contribute to faster corrosion.

b) Ageing of plastic materials and rubbers - which makes it partly or fully impossible to use the above materials in design work. Atmospheric ageing is a complex of mostly irreversible material changes, induced by atmospheric elements. The result is changes in size and dimension, outer appearance, and particularly in a considerable decline in the initial mechanical properties.

c) Declining power of combustion engines - due to high temperatures of the surrounding air. As a result of fuel heating, changes in physical properties of engine fuel take place, and the weight of fuel being carried towards the engine becomes less. In addition, increased air temperature, and/or higher atmospheric pressure decrease the unit weight of the air admitted by the engine. The two phenomena would reduce the potential design engine power.

d) High failure rate of electrical equipment, particularly of electric motors, contactors, and of other electric instruments, which are exceedingly sensitive to extreme climatic factors. An accessory phenomenon is changes in the properties of the electro-insulation materials, corrosion of contacts and conductors, as well as a number of other degradation effects.

e) Reduced efficiency of radiators, coolers, and air filters in very dust-rich air of the arid zones, and also during operations carried out in tall-grass stands. Consequently, the radiators and coolers should be designed with regard to higher temperature of the incoming air.

f) Increased wear of movable mechanisms and parts like bearings, piston engines, fuel injectors. This increased wear and the resulting shorter life is predominantly of an abrasive nature, being most pronounced again in the arid regions.

g) The biodeterioration is any change (degradation) of the non-metallic materials which is caused by living organism lige rodents, moulds, termites, and the like.

2. The Effect of Production and Crop Management Factors

Considerable variation of the production and crop growing and management conditions, as well as the existing difference in this respect between those of the developing and equipment manufacturing countries may result in machine failures even when the machine in question is well designed and made to be weather-resistant. It has been a lack of this knowledge, and the resulting non-observance of different production and crop management conditions that has led to so many complaints when exporting tractors to Ghana and Japan. The factors under consideration may be divided tentatively in three groups:

a) The production factors characterizing the general conditions under which the machine is to be operated, e.g. the pattern of land use and field size, different length of the work-day, possibilities of a full engine power and/or machine operating width expoitation, etc.

b) The soil conditions - it is well known that tropical soils modified annualy by the alternating rainy and dry periods are difficult to work, their clay and silt content being much higher. The soil resistance not infrequently exceeds the value 1 kp/cm^2, and as a result of a higher soil resistance, the tractor horse-power must be greater too.

c) The crop management factors, i.e. the effect of the crop grown, as well as the local cultural practices have a bearing particularly on the harvesting equipment, threshers, cultivators and similar machines used throughout the growing season. Mechanical properties and morphological features of a certain crop grown in the temperate zone differ from those of the same crop grown in a tropical country, in addition, sometimes certain crops must be harvested the mechanical properties of which (their stalks, seeds, etc.) are not known.

3. Effect of machine Operation skills

It is a well known fact that skillful machine operation exercises a significant effect on the efficiency, failure rate, and life of farm tractor and the attached machine aggregates. It is also well known that the development in design of modern farm machinery has been faster than the training of new tractor and other operators even in the developed countries, and according to the available projections, sophisticated agricultural equipment of the so-called second and third generations will have to be operated by driver-mechanics with a secondary school level education.

In this respect, the situation regarding training and instruction of future machine operators in the developing countries is much less satisfactory. The in-training of local machine operators in the developing countries has been too one-sided so far, consisting in fact in the basic schooling of tractor operation and repair servicing.

It should be also taken into account that the climate and nutritional situation in the developing countries are not favourable from the view-point of labour performance, and do not permit intensive work throughout the work-day, or a full attention to the machine operating. A number of scientific papers and my own experience have shown that the technical know-how of the machine operators is below that of the developed countries. This is inter alia due also to unsatisfactory general polytechnic education, as well as due to the present methods of tractor-operator training.

It is the above experience that must be considered in our decision-making on the types and machine designs that would be best under specific operating conditions from the view-point of available manpower skills.

4. Effect of Transport and Storage

It does not differ in the case of farm equipment from that affecting other mechanical engineering products. Of greater significance in this respects is however the storage or housing of farm machinery during the off-season to sever particularly the impact of extreme climatic conditions.

It is more than evident from the examples and reasons given above that the farm tractors and equipment exported to the developing countries must be factory-designed and made with regard to their environmental conditions, and any later and additional adjustment and improvement would only be a partial answer to the problem and, would in most cases make the cost incurred unbearable.

Problems of Agricultural Mechanization in Tropical and Subtropical Regions

by J. Oubrecht

The major centres of farm machinery and equipment manufacturing are in Europe, North America, Japan, and Australia, respectively, while the regions featuring most extensive untapped farmland resources and low population densities (Africa, South America) have been increasingly lagging in the production of high-performance modern farming equipment without which any exploitation and improvement of the so far unfarmed lands is impossible. The marketable crop production in a number of African and South American countries is so low that it fails to meet the requirements of the constantly growing urban populations in the areas in question. The progress in agricultural mechanization in the developing countries has been impeded by surplus cheap manpower, and, particularly by a critical shortage of skilled labour. Nevertheless, the major world firms manufacturing farm tractors, equipment, and implements have been conquering more and more the machinery market of the developing countries despite of the limitations given above, and consequently the numbers of skilled equipment operators, maintenance and service technicians has been on a permanent increase. This is all the more important since the developing countries should be in a position of metting their higher food requirements from their own domestic sources.

A considerable decline in the rural population e.g. in Iraq had a beneficial effect on productivity, yet the total farm output has not risen, an unwelcome fact that has widened the existing gap between the food supply and demand. The food import possibilities are limited, in most instances by lack of foreign exchange. On the other hand the crop production in Europe has attained high levels of intensity, and there has been still a sufficient scope for further productivity and intensity increases. It is thus necessary for the advanced countries to help the developing countries in the field of food production, and the cuts in arms spending should be reflected in expanded and improved training of agricultural specialists for the developing countries, and in the introduction of modern land use and farming methods in the countries of the Third World.

Mechanization of the cultural practices and operations has attained significant levels for those crops that are being grown also in the advanced countries, i.e. often for those crops having no substantial bearing on improvement of the food and nutritional situation of the developing nations. For instance, the management and harvesting of the following major crops grown in tropical and subtropical regions has been fully mechanized: maize, rice, cotton, sugar cane, grondnuts, etc.

The engineering plants manufacturing farm machinery and equipment do not find any difficulties whatsoever to turn out harvesters handling the sorghum, lovegrass, and other crops essential in human nutrition in the developing world. However, the usefulness of sophisticated farm machinery in the developing countries of the tropics and subtropics depends to a large measure on the designer's and maker's skills to take fully into consideration the fact that the soil resistance tends to be here generally higher: in the rainy season both the soil moisture and air relative humidity are high, while during the dry season the reverse takes place, and the air dust and dust storms make the equipment operation extremely difficult. Had the equipment maker taken into consideration the factors mentioned above in the machine design, his product is suitable and operational also under difficult extreme climatic conditions. Also the farm equipment manufactured in the C.M.E.A. countries is fairly well applicable in the tropical and subtropical regions, yet it is imperative to test the to-be-exported machines first under the environmental conditions they are supposed to operate, and to supply and use them on a large scale only after the necessary tropicalization and adjustment. It would be thus irresponsible to sell a certain farm machine to a developing country with a full knowledge of the respective operating conditions, and being aware of the fact that it had not been tested earlier in the very same conditions or simulated ones. It should be also realized that the efficiency and performance of the farm equipment sets in tropical and subtropical regions will be less for the following reasons:

(1) The power source output will decrease when operating in an atmosphere of high surrounding air humidity.

(2) It is a well-known fact that the equipment operators are in general not skilled enough, this affecting adversely the machine productivity.

(3) Also the soil and climatic features of tropical and subtropical regions have a substantial impact on the equipment performance.

(4) The roads are mostly unsatisfactory, and the field topography would exhibit very poor levelling.

The soil tillage machinery (and some other too) have been designed in most cases for operation in the temperate zone and there are a few instances only of the farm equipment designed specifically for a particular crop or in consideration of the tropical and subtropical climates and environments.

It must be emphasized that any exports of the available farm equipment to the developing countries are not feasible without it being first pre-production tested on the very spot. It is also essential to develop well ahead of time the machine servicing facilities staffed by experienced repairmen. Adequate technology and sufficient space of the service workshops are the necessary prerequisites for prompt and perfect repairwork and remedy of the machine failures. Such service centres are to serve at the same time as training centres for operators of sophisticated farm equipment sets. The objective is not only to train skilled equipment operators not only from the view-point technology, but also with regard to optimum cultural practices, i.e. to teach them judging the quality of the work performed. In this era of comprehensive farm operations mechanization is the role of the tractor and similar complex equipment operator very taxing, and can be done by very capable persons only.

My own personal experience has convinced me that the ultimate effect of the field operations is due - apart from an efficient equipment and trained operator - also to a skilled manager who should provide on the remote locations

(1) Good though temporary housing,

(2) Sufficient supply of good food and potable water,

(3) First aid kit, etc.,

(4) Most important spare parts that may be mounted in the field,

(5) Machine maintenance tools, etc.

The absolutely largest differences in the equipment productivity between the temperate and tropical zones are encountered in the virgin land tillage for drilling; these are due to considerable soil resistance, obstructing stones etc., and frequently ploughs fail if they had not been designed with regard to the specific nature of tropical and subtropical environment. The ploughs manufactured in the C.M.E.A. countries would be designed mostly for the soil resistance up to 9,000 kp/sq.m., yet that encountered in the virgin land tillage operations might be often considerably higher.

The experiences of the Farm Machinery Industry in the G.D.R. in Export and Utilization of Agricultural Machinery in Developing Countries

by R. Hofmann

When we have a look at our 25 years successful development of the G.D.R. this year, then we recognize also our success by the achievement in the development of agricultural and foodstuff technique.

That way was marked by two significant appearances in our social life. Those are: The development of a modern socialist agriculture on the base of a close unit of the workers' class with the working farmers and the continuous improvement of friendly cooperation with the Soviet Union and other countries of the C.M.E.A.[x] They have had a remarkable influence on the development of the production and market relations of agricultural and foodstuff engineering, and on the other hand the technical and economical politics of our industrial branch has given impulses for the quick rise of agricultural production and the successful scientific-technical and economic cooperation of the countries of the C.M.E.A.

Allow me to give some details, concerning the industrial branch of agricultural and food-stuff technique of the G.D.R., for the explanation of our attitude in the relations to the developing countries and for the understanding of our possibilities for an improvement of these relations.

1. The industrial branch of agricultural and food-stuff technique in the G.D.R.

When in 1948 the nationally owned branch of farm machinery industry was founded, there were only a few very small enterprises, which began the production out of the ruins of the war with the help of Soviet friends.

---

[x](Council for Mutual Economic Aid).

The Soviet help and the experience of the workers and technicians of so traditional farm machinery enterprises like Petkus Wutha and Saxonia Bernburg, were the base on which a rapid development of farm machinery industry was built up. In 1949 there were first tractors and trucks from the Soviet Union which formed the base for the machine-hiring-stations (MAS) and therefore for the mechanization of agriculture, but then, in 1952, we had already the first big machines from the G.D.R. production, like milking machines, tractors RS 09 and chaintractors, which, together with the tractors and combine harvesters of the type SK4 from the Soviet Union began to change the face of agriculture.

In 1952 the first steps for the specialization and concentration of production were done by the unification of 5 smaller farm machinery enterprises to the VEB Kombinat Fortschritt Neustadt.

This process was significantly influenced and accelerated by the foundation of the first cooperative farms in Worin and Merxleben.

From this time on the most important steps of development in farm machinery industry have always been coordinated with the fixation of decisive steps of development in socialist agriculture.

In this sense the directions of the 8th party congress of the SED[x] for the 5 years plan for the development of political economy in the G.D.R. from 1971 - 75 among others set out:
"Under the conditions of the G.D.R. the further increase in the production of agricultural products has to be accomplished by socialist intensification, that means among others, by the use of chemicals and complex mechanization of crop and animal production." (1)
This process of the socialist intensification of agriculture is a task, which will not only be solved within a national frame, but also within a frame of socialist economic integration, which is proceeding quickly.
In the special C.M.E.A.-section for agricultural machines and tractors, a series of machine systems was already acknowledged on an international basis by 1958.

---

[x]Socialist Unity Party of the G.D.R.

The complex programme of the C.M.E.A. (1971) also comprises the common aims and the cooperation of agricultural and foodstuff machinery industry of the countries of the C.M.E.A.

Let me explain by some figures and statements the development and importance of our industrial branch:

a)

Year	Production (in million marks)	Export (in million marks)	Export to developing countries (in million marks)
1950	55	1.5	-
1973	3,000	1,400.0	30

b) In our industrial branch about 60,00C workers, employees, engineers and scientists are employed. The production is placed to 50% at the disposal of the agricultural and foodstuff economy of the G.D.R. and exported to 50%, from that to the largest part of the C.M.E.A.-countries, on the basis of socialist economic integration.

c) The production programme of the industrial branch comprises in its 6 combined works, and those are:

VEB Kombinat Fortschritt Neustadt,
VEB Weimar-Kombinat, Weimar,
VEB Kombinat NAGEMA Dresden,
VEB Kombinat Ascobloc Dresden,
VEB Kombinat Impulsa Elsterwerda,
VEB Kombinat agrotechnic Leipzig

a large assortment of agricultural and foodstuff machines with the following complexes:

- Machines for the kernel and fodder crops production, the machines and equipments for the mill industry inclusively;
- machines for the potatoe and sugar beet production;
- machines for the milk production and processing;
- machines and equipment for industrial methods of animal husbandry;
- machines and equipment for the processing of meat;
- machines and equipment for the pastry, beverages and sweets industry;

- machines and plants for the social food production and the potatoe refinement;
- machines and plants for the packing and weighing technique.

The international co-operation on the base of the complex programme of the C.M.E.A. will lead to a continuously closer communication of the industrial branch of the agricultural and food-stuffs machinery too.

So our industrial branch will be concentrated upon main products of the above mentioned complexes and partial sections of the machine systems, and the production assortment will be limited to about 400 final products within the frame of specialization and concentration of production in the following years.

The balanced conception of the C.M.E.A. countries for an integrated production of the necessary machines and machine systems for the intensification process in agriculture places us in the position to furnish the agricultural and food-stuff economy of our countries with the most modern techniques and besides to improve and enlarge our relations to the agricultural and food-stuff economy in the developing countries.

In the following section of my expositions let me briefly explain our principal idea in foreign trade policy concerning the relations to the developing countries. In the already cited direction on the 8th party congress of the SED the following principle for it is formulated: "The foreign trade and the scientific-technical relations of the G.D.R. to the developing countries have to be further developed. They are intended to contribute to the improvement of the industrialization and export power of the developing countries and to the enlargement of a stable, long-term reception of economically important goods from these countries."(2)

2. The principal ideas of the foreign trade policy of the G.D.R.

The foreign trade policy of the G.D.R. intends to give a contribution to the overcoming of the economic disadvantage in the developing countries and to enlarge the economic and scientific-technical relations continuously.

We have in view, that such economic relations help to support these countries in their struggle for political and economical independance against imperialist states, to stabilize the economy of the developing countries and to make the advantages more effective of the international labour sharing based upon equality of status and mutual advantage.

In order to use the possibilities and necessities of such a labour sharing fully, the G.D.R. aims to conclude long-term trade agreements and contracts. Such agreements include the readiness of the G.D.R. to conclude long-term contracts with the developing countries in order to improve the mutual exchange of goods in the interest of both. Here we take into consideration the natural conditions of production and the export-structure of the developing countries, as well as the traditional and new export goods, in order to contribute to the development of industry and agriculture in those countries. In their foreign trade policy the G.D.R. views that such agreements contribute to

- a continuous and rapid development of the exchange of goods, based upon equal rights and mutual advantage,
- an improvement of the co-operation and support in the composition of the economic programmes and significant economical complexes in the developing countries by experts of the G.D.R.,
- a close component of the economic relations in the scientific-technical co-operation as well as the training of specialists.

3. The realization of the foreign trade political base of the G.D.R. by the industrial branch of agricultural and food-stuffs technique

This base of foreign trade politics of our state, orientated towards the effective support of economy in the developing countries also determines the practical activity of our industrial branch in the relations to the developing countries. There we are conscious, that especially our help as a producer of farm and food-stuff machinery is useful for one of the most important sectors of political economy, viz agriculture, which supplies a decisive part of the national income in the developing countries.

Permit me to explain to you briefly the most important methods of co-operation of our industrial branch with the agricultural partners in the developing countries.

3.1. Export of farm and food-stuff machinery for the mutual advantage

Our principal idea is, and this has been proved since years for the advantage of the producer and the user, that no machine is exported which we did not prove in all respects regarding its application for the utilization under the provided conditions.

The judgement for the application is formed

- by the testing of single machines or systems at the intended place of utilization under natural conditions,
- by the positive results of utilization in the own agriculture or in agricultural enterprises of other countries with such machines, where the utilization admits analogical conclusions,
- by the testing of the machines in our research and testing stations by the creation of simultanous situations or by artificially made conditions.

All those special tests of course precide an intensive testing in the research and testing stations of our plants and the testing stations of our agriculture.

A further principal idea is that we consult and agree upon the conditions for the utilization of these machines together with our partners in the developing countries, according to a close, long-term contact, already before the conclusion of the contract concerning the supply of machines.

Those consultations and agreements comprise

- the most suitable technology for the utilization
- the classification in the system of the present techniques,
- the necessary qualification of the operators and the personal for maintenance,
- the security of spare-part-supply and base of maintenance,
- the service for utilization by the producer.

3.2. Scientific-technical co-operation

Within the frame of commercial co-operation stretching over many years we have developed the scientific-technical co-operation as an integral component of our foreign trade relations. Thereby we are guided by the 3 main aims of the scientific-technical co-operation, as they were formulated by socialist countries at the first conference of the United Nations for trade and development:

- the scientific-technical co-operation must be directed upon the stabilisation of the political and economical independence of the developing countries,
- the scientific-technical co-operation could improve the rise of national economy on the base of industrialization, of the development of agriculture, of the diversification of production, of the foreign trade of the developing countries as well,
- the scientific-technical co-operation with developing countries should underlie to no political, economical, military and other conditions which should not be acceptable to the developing countries.

  How have these aims of the scientific-technical co-operation been realised in practice by our industrial branch?

  Our activities in this field can be comprehended in the following three complexes:

- A decisive place in our scientific-technical co-operation takes the complex scientific support in exporting our machinery and equipments.

Thereby this support begins with the technical and technological consultation on the base of present conditions in the country where the machinery should be used. Typical examples of our working method are the preparations for the export of rice- and wheat-mills in the A.R.E. and the Irakian Republic, for the supply of milking plants in the A.R.E. and the Syrian Republic, the export of Combine harvesters to India and to the A.R.E. or the export of farm machinery to Algeria and to other developing countries.

The second step in the frame of this support consists in the dispatch of experts of the manufacturer, who guide and supervise the setting up of the equipment, the machines and installations and instruct the machine-and repair personnel in the first phase or in the main season.

Constituent part of these measures is the training of the operators and repairing personnel in the factory of origin or on the agricultural farms of the G.D.R.

The cited measures furthermore include on extension service about planning and storing of spare-parts and organization of maintenance.

The dispatch of experts from the G.D.R. for solving certain problems characterizes a further complex of out scientific-technical cooperation. In order to carry out these problems we have veryfied relations to other institutions in the G.D.R., such as to the Institute of Tropical Agriculture and Veterinary Science which is holding this summer course, to agricultural faculties, to plants of the agricultural- and food-stuff-economy or to the Ministry of Agriculture. With the help of these institutions we are in a position, besides the possibilities of the enterprises, to dispatch experts for the solving of special problems to our partners in the developing countries, respectively to employ experts for the training of our foreign partners.

Thus, G.D.R. agricultural experts among others, to cite only a few typical examples, introduces the complex use of combine harvesters in the A.R.E., experts of the mill-industry trained the leading and technical personnel for mills which were erected by us or experts of the Institute of Tropical Agriculture and Veterinary Science worked in the testing of our machines.

The third complex of our scientific-technical co-operations can be marked by the term "industrial co-operation and bestowal of licence".

In this form of co-operation it becomes evident that we fully explore the possibilities of the industrial development of the developing countries.

Thus with the conclusion of comprehensive contracts about the supply and construction of rice- and wheat-mills to the A.R.E., we agreed on the production of about 30 % of the value of the mill-equipment in the country of the partner. Special significance we ascribe to the possibility of the bestowal of licences for the production of certain farm machinery. Thus the licence for the production of ploughs was bestowed upon an Egyptian enterprise.

At the same time we are in discussions with some other countries for instance with India.

Before I finish with my remarks about the principles of our relations to the developing countries, I should like to say that our price and credit policy too is orientated to help according to our possibilities. So, a lot of the concluded contracts with the developing countries are based upon government - or firm-credits, which offer favourable conditions of payment to our partners.

4. Possibilities of the industrial branch for the supply of machines and machinery to the developing countries

I should like now, having in view the experience of the last years and the future development of our industrial branch, to explain some thoughts about the possibilities of the supply of agricultural and food-stuff machinery. Thereby we always have in view, how, under our conditions, we can contribute to the solving of the mechanization problems in the developing countries.

In the beginning of my explainations I have already pointed to the fact, that our production profile is determined to the gradual change to agro-industrial technology in the agriculture of the G.D.R. and the agriculture of the C.M.E.A. states.

Gerhard Grüneberg, member of the ZK of the SED, formulated in a speech, made in the beginning of this year, this fact as follows: "For the socialist intensivation of agricultural production and its change to agro-industrial technology it is typical that the composition of agriculture in all fields of political economy will be closer. First of all it is about the composition of agriculture with the branches of industry, producing means of production, and the branches of storing, manufacturing and the market for agricultural products and raw-materials. The consequence results in the fact, that it is no longer sufficient, to look upon the co-operative farmer and the agricultural labourer as the only producers of food-stuffs in agriculture. Food-stuffs are produced by all working people under the conditions, based on agro-industrial-technology in agriculture with all its composed branches of political economy". (2)

In this process our industrial branch plays a decisive part. It makes products available - this includes the completion of assortment in the frame of the C.M.E.A. countries and also from capitalist industrial countries which guarantee a mechanized cultivation and harvest of the main crops, a partly and fully automaticalized storing and processing, inclusive the packing or food-preparation in largescale kitchens.

This complexity also offers favourable prospects to our partners from the developing countries for the preparation and carrying out of measures for mechanization, which contribute to achive a higher, qualitative better production of food-stuffs, for the own demand or for export.

In this sense, for instance, consultations and discussions with the General Organization for Utilization and Development of Reclaimed Land of the A.R.E. take place at present, which intend to construct together an agro-industrial complex in a sector of reclaimed land area to produce milk and meat. Such a project shall include: The mechanized cultivation, transport, storing and the mechanized production of the necessary fodder, the mechanization of cattle production inclusive the feeding and dung removal, the mechanization of the milking-process, the complete technical equipment for the milk industry to milk for drinking, cheese and butter, and the complete technical equipment for the butchering and meat processing.

Though the development of agriculture in the developing countries in general shows a low step so far, tendencies in some countries already become visible, that modern agricultural farms are formed on the base of the state sector and, as that becomes evident in the A.R.E. - even whole agricultural areas strive to the development of agro-industrial-complexes.

Our production assortment offers favourable prospects, especially to the state sector in agriculture, which represents the economic base for the revolutionary, democratic politics in developing countries, which have been going on the noncapitalist way of development, to grant generous support. Similar favourable possibilities for the use of our products are given by the development of agricultural cooperatives.

Considering our standard products and making use of the knowledge about the special conditions in developing countries, and using the testing results, we have carried out changes in the construction of some machine elements (as for instance the fitting of parts protected against tropical (Limate), and in the modification of some machine parts as well (as for instance sack-installation at the combine harvester instead of the corn bunker).

Furthermore, basing on our standard machines, we carried out special cevelopments for the enlargement of utilization of our products. So, for instance, special cutting parts for the combine harvester of the type E 512 were made for the maize and rice harvesting.

The continuous further development of our equipment construction places us in the position to project and carry out complete objects for the developing countries. Thereby the inclusions of local production and the incorporation of products of other C.M.E.A. countries for the security of the complete production assortments is getting always more importance.

5. Export and utilization of G.D.R. products

The results and experiences in the test station El-Ghasair (A.R.E.) demonstrate clearly, that we intend to give real help for the development of agriculture in developing countries. As result of the activity of the test station we have recommended among others to the Egyptian agriculture ways,

- how to improve soil fertility by tillage,
- how to harvest the main crops wheat, maize and rice and a varities of other crops with the lowest crop losses and at an agrotechnical favourable date by the use of the combine harvester
- how to accomplish the fodder-crop cultivation and harvesting fully mechanized,
- how to achieve a production rise by proper tillage mechanized sowing and plant husbandry and finally also
- how to enlarge the field of use of modern techniques by new methods of irrigation.

The introduction of milking equipments in the state sector of the Egyptian agriculture is based on a close co-operation between experts of the VEB

Kombinat Impulsa and the General Organization for Meat and Milk of the A.R.E. This co-operation began with the choice of the most practical techniques and technology and has been continued after the start of the first enterprises, to make use of the experiences for the equipment of further plants.

With the employment of Impulsa milking-plants by this close scientific-technical co-operation of experts from both countries it has been achieved,

- that all big state farms for cattle production carry out the milking process by machine
- that for the first time in big farms of buffalo breeding the use of milking-machines could be carried out successfully,
- that in continuation of the started co-operation on the field of milking techniques further sectors of animal production have been included in the mechanization process.

With the delivery of wheat-mills to the Syrian Republic we have placed our partners in the position to achieve already now 25 % of the necessary milling-capacity of Syria with a high quality. In 1978 about 80 % of the necessary milling-capacity will be guaranteed by modern big mills from our industrial branch.

Also in the A.R.E. about 90 % of the flour production in Egypt with mill-machines from the G.D.R. has been achieved after the exploitation of the supplied big mills and after the conclusion of the reconstruction measures in the present mills.

These projects had been erected in work-sharing co-operation and guarantees a production of food-stuffs with highest quality on the base of modern techniques and technology and a solidly trained personnel.

The significance in modern techniques in the production of food-stuffs of high quality is evident, among others also in connection with the export of our rice-mills. With the start of the first big rice-mill in the A.R.E. a rapid rise of export quality of rice has been achieved. The completion of all delivered equipments will be forced along, because it should be achieved, that in 1975, the whole paddy, intended for the export, will be peeled in our mills, in order to improve the capability of competition and to rise the export gains significantly.

6. Final remarks

To summ it up the perceptions, we can make in the relations with the developing countries, as follows:

The export and the utilization of our products, together with the connected thereby scientific-technical co-operation has contributed

- to introduce yield-increasing technologies
- to secure harvesting storing and processing of important food-stuffs with low losses and high exploitation,
- to rise the quality of important food-stuffs,
- to contribute to the development of industrial production and to the rising of export gains in those countries,
- to improve the technical qualification of leading personnel, of workers and farmers.

Let me finally once more mention the report of Gerhard Grüneberg, member of the politbureau of the SED, about problems of agrarian politics. In connection with the representation of the increasing disproportions, between food-stuff production and the development of the world's population and the thereby caused reaction on the world market, he expresses among others: "For the G.D.R. then involves the problem to secure the needs of basical food-stuffs alone, as far as possible, and also to secure the want of most necessary raw-materials alone, in co-operation and work-sharing with the socialist countries, especially with the Soviet Union, and, according to its possibilities, to contribute to the nutrition problems in the underdeveloped countries and young national states." (4)

With the basical orientation I stress my contemplations, that the industrial branch of agricultural- and food-stuffs-techniques indirectly by its contribution for the intensifiation of agricultural production in the G.D.R., as well as directly by its export of agricultural- and food-stuffs-machines to the developing countries fulfils a real political and economical help for the construction of national economy in the young national states.

References

1 Direktive des VIII. Parteitages der SED zum Fünfjahrplan für die Entwicklung der Volkswirtschaft der DDR 1971-75, Berlin 1971, p. 94.

2 op. cit., p. 112.

3 Grüneberg, G., Die gegenwärtigen Aufgaben bei der weiteren Verwirklichung der vom VIII. Parteitag beschlossenen Agrarpolitik der SED. Berlin 1974, p. 16.

4 op. cit., p. 9.

Technical Training for Students of Tropical and Subtropical Agriculture at the Karl-Marx-University, Institute for Tropical Agriculture and Veterinary Science, Leipzig

by R. Neumann

1. The Aim of the Course

In the education and training for students, consideration must be given to the function which they are to perform after graduation. It has been our experience that, after their return to their home countries, the students usually take up top positions in the agricultural ministries, scientific institutions and farms or on important projects, and thus the technical training courses must be structured so as to provide students with the technical knowledge about tractors and agricultural machines which are indispensable to successful management in agriculture. This aim is, of course, set out in the course plan, and in the general curriculum [1], section "Education and Training - Aim", it is stated that the students must be trained to work scientifically, to evaluate and exploit scientific discoveries and methods and to apply them creatively within the framework of the prevailing conditions in their home countries.

They must be able of introducing new production techniques into the agricultural set-up of their countries whilst at the same time giving due consideration to the relationship between natural-scientific, technological, social and political conditions. The students are required to study and make use of the scientific and political experience gathered by the G.D.R. and the other socialist countries.

The experience which has been gained from the almost totally mechanised set-up of agriculture in the G.D.R., from the training of agricultural personnel and the discoveries and experience which our institute has been able to make in the field of agricultural mechanisation in tropical and subtropical countries provide a solid basis for the training of agricultural specialists for and from the developing countries.

2. Technical Training and Education

2.1. Technical Subjects

The general aim for education and training also forms the basis for the training programme in farm machinery. It is our job to acquaint the students with the physico-technical principles and with the agricultural machines and tractors presently in the use in production. We fulfil this task by dealing with the three subject groups:

Technical Physics	40 h = 1.10 %
Basics for Farm Machinery incl. Driving Course	135 h = 3.75 %
Machinery for Crop and Animal Husbandry incl. Course on Combined Harvester	150 h = 4.16 % (25 h)
	325 h = 9.00 %
	3,600 h = 100.00 %

The total time available for the general training course is 3,600 h (without M.Sc. course). 325 h, or 9.00 % of the course is devoted to training for farm machinery. Thus, it can be seen that the two subjects "Technical Physics" and "Basics for Farm Machinery", occupy a relatively large portion of the course.

This broad basic training course is, in our opinion, necessary as it makes it possible to deal with fundamental machine parts and sub-assemblies which are often to be found in agricultural machines. In this way the students are equipped with a knowledge which will remain valid over a relatively long period. The machinery of crop and animal husbandry, however, is typified by a more rapid process of change and development.

A considerable part of the training therefore consists in providing the student with the knowledge, know-how and ability he will need when dealing with technical processes, tractors and agricultural machines. This is designed to enable him to decide which machines are most appropriate for the job in hand so that he can work out suitable technological programmes and direct production.

The material to be dealt with during the course in distributed over the various terms in the way which best suits the students' development.

Technical physics	2nd term
Basics for farm machinery	2nd and 3rd term
Machinery for Crop Husbandry	4th and 5th term
Machinery for Animal Husbandry	7th term

The individual subjects are dealt with in the following passages.

2.2. Technical Physics

The students at out institute come from different countries with different educational systems and thus their basic knowledge of the natural sciences and the standard of their general training have reached different levels by the time they arrive in Leipzig. We must, of course, take this fact into consideration and therefore we try to overcome these differences at the very beginning by means of a course of lectures dealing with the basics of natural science.

This process of placing the students on an equal footing is intended exclusively as one step towards producing graduates who live up to and fulfil the aims of the course as set out above - it is not an end in itself. Thus we use the term "technical physics" to mean a subject which equips the students with a basic general knowledge which will serve as a basis for their training in farm machinery for crop and animal husbandry.

Many of the physical processes and laws can be seen in a very practical form in the processes connected with agricultural machinery and thus can be applied and expolited in this field. We attach great importance to a practical approach to teaching the physical laws and processes as it enables the students to grasp the concepts involved much more easily than would a formal approach. Apart from increasing the rate of teaching success, this method also provides students with knowledge which will stand them in good stead over along period. The content of this part of the course is therefore determined according to the following rules:

Those theories are to be taught which are important in agriculture. The essence and substance of physics and physical laws are to be taught with the emphasis on their practical applicability. The students should be able to recognise aspects of physics and the physical laws as they appear in practical situations. The creation of a fundamental body of knowledge in physics and the natural sciences is the starting point from which the subjects

- basics for farm machinery
- machinery for crop and animal husbandry

can be developed in later sections of the course.

We can illustrate the dependence of the various subject groups by means of an example:
The lecture on "The Mechanics of Liquid Bodies" deals with basic aspects and laws of hydrostatics. Special emphasis is placed on those aspects which are essential to an understanding of hydraulic processes and in the basic course on farm machinery this subject is further developed by the section on hydraulic units for tractors. The subject group "Machinery for Crop Husbandry" takes up this theme and deals with the hydraulics of agricultural machines and the possible ranges of application.

The subject "technical physics" ist followed up by "Basics of Agricultural Technology".

2.3. Basics for Farm Machinery

This subject is designed to provide the student with a general training in farm machinery which he will need if he is to be successful in his future work. In our opinion, the following areas should be dealt with:

- Materials in use in agriculture
- Machine parts
- Internal combustion engines and tractors
- Mechanical handling devices
- Pumps and blowers
- Maintenance
- Aircraft in agriculture
- Automation technics
- Driving course

The largest part of this course is given over to internal combustion engines and tractors (40 h 30 % of the teaching time). This subject is dealt with in great detail as combustion engines are of considerable importance to the farmer in their role as power sources for stationary units, self-propelled agricultural machines and tractors. The tractor remains the most suitable source of power for agricultural purposes and thus has a major role to play. It will certainly maintain its key position over the next few decades despite the appearance of self-propelled units. The students should therefore be closely acquainted with the structure, function and interaction of the tractor's various sub-assemblies with special regard to their roles when in action in agriculture. The student must be able to evaluate the potential of a tractor so that an efficient rate of utilization can be ensured when using aggregate units.

When dealing with materials and machine parts, special attention is paid to maintenance as it is the farmer himself who has the greatest say in the maintenance of tractors and agricultural machinery. Proper care of the most varied sub-assemblies is essential if the machinery is to function correctly and it has been our experience that the success of agricultural work in the developing countries linked with the level of maintenance.

The importance of maintenance is demonstrated to the students in a practical way during a visit to a servicing station for agricultural machinery.

The farmer must also be acquainted with mechanical handling devices, pumps and blowers not only because they are important in handling operations as independent units but also because they are very often employed as sub-assemblies with various other machines.

Aircraft are being used to an increasing extent in agriculture for jobs such as pest and weed control (plant protection). For this reason, the student must have at least a basic knowledge of agricultural aircraft and the equipment used in conjunction with them.

Automation technics is dealt with in two courses of five hours each which cover control techniques.

The driving course completes the basic training course. Here, the student learns to drive a tractor and takes the appropriate driving test - according to G.D.R. traffic regulations. We consider the ability to control and drive a tractor essential as it gives the students a better idea of the jobs for which tractors can be used and of the potential of the vehicle in various situations. Apart from this, driving ability is extremely useful in exercises and practical training during the course itself.

After completing the basic course and passing the driving test, the student is ready to go on to the specialised trainig in farm machinery, a course on crop and animal husbandry.

2.4. Machinery for Crop and Animal Husbandry

Machines and equipment used in agriculture are of primary importance to the farmer. The significance of farm machinery will certainly increase as time goes on and even today agricultural production depends primarily on machinery; for this reason it is absolutely essential that our students are equipped with sufficient knowledge of the farm machinery.

The course is designed so as to give the students a survey over the complex field of processes in tropical and sub-tropical agriculture.

Although the material is dealt with systematically, the following table is arranged according to the amount of teaching time devoted to each section:

1. Harvesting machines and equipment for threshing crops	23 h = 15 • 3 %
2. Machines and equipment for tillage	15 h = 10 • 0 %
3. Machines and equipment for plant protection	10 h = 6 • 6 %
4. Machines and equipment for perennial crops (and woods)	10 h = 6 • 6 %
5. Harvesting machines and equipment for forage	8 h = 5 • 3 %
6. Machines and equipment for seed cleaning and grading	7 h = 4 • 7 %

7.	Machines and equipment for sugar bearing crops	6 h = 4 · 0 %
8.	Machines and equipment for seeding and cropping	6 h = 4 · 0 %
9.	Machines and equipment for root crops and ground-nuts	6 h = 4 · 0 %
10.	Machines and equipment for spreading fertilizer	5 h = 3 · 3 %
11.	Machines and equipment for fibre crops	5 h = 3 · 3 %
12.	Machines and equipment for field vegetables.	5 h = 3 · 3 %
		106 h = 70 · 6 %
13.	Machines and implement for the production and processing of animal products	6 h = 4 · 0 %
14.	Machines and plant for the preparation and processing of fodder	5 h = 3 · 3 %
15.	Machines and plant for fodder distribution	3 h = 2 · 0 %
16.	Machines and plant for processing waste products from animal production	1 h = 0 · 7 %
		15 h = 10 · 0 %
	Final colloquium	4 h = 2 · 7 %
	Course on combined harvester	25 h = 16 · 7 %
		150 h = 100 %

We believe that this structure takes appropriate account of the individual subjects to be dealt with.

Primary importance is given to machinery used in harvesting threshing crops because many other factors are directly dependent on the quality of the harvest (reduction in losses, target dates and deadlines in agricultural technology, storage and processing).

In the developing countries these are important major crops.

The amount of time available, however, allows variation on the above plan so that - depending on the students taking part in the course - one or other of the subjects can be dealt with more thoroughly (eg. machines for harvèsting fibre-crops). The time devoted to machines and plant in animal husbandry will definitely be extended in future as new processes and methods involving modern machines and equipment are constantly being introduced into the field.

The theoretical side of the course on crop and animal husbandry is divided into

- Tasks of the machines (working principles)
- Structure of the machines
- Functioning of the machines (functional principles)
- Characteristic data and parameters, performance and efficiency
- Modification of the machines, equipment and plant.

The systematic ordering of subjects here has proved very favourable as the student gets to know the machines and their performance and modifications in gradual steps according to the site and the job in hand. The knowledge which the student gains here about the machines themselves serves as a basis for the following course on technology which consists of lectures on sub-assemblies or machines and of practical exercises such as:

- Exercise on tillage and fertilization
- Exercise with machines for plant protection
- Exercise on cleaning and grading of seeds
- Exercise with milking machines
- Exercise on milking plant.

The training involving crop production machines is completed by a course on combined harvester which takes up 25 hours and during which the students are acquainted with important machines of equipment such as combined harvester, straw balers and forage cutters.

3. Pedagogical aspects of the form of the lessons

The student should be introduced to material which is completely new to him and this should be done in such a way as to capture his interest and firmly anchor the knowledge in his mind. The teacher must try at all times to achieve a unity or balance between the level of achievement and enthusiasm of the students and the quality of presentation and development. This presents considerable difficulties and means that the successful teacher must be in constant contact with his students. The right balance between the various forms of lesson must be struck i.e. lectures, seminars, exercises and colloquia. The relation between lectures, on the one hand, and seminars exercises on the other is as follows:

Technical physics	3 : 1
Basics of farm machinery	2 : 1
Machinery for crop and animal husbandry	2 : 1

In our lessons we use audio-visual methods such as slides, charts and illustrations, films, overhead projectors etc. At the moment our section has over thousand slodes of its own. It has been our experience that many of the students have difficulty in grasping the technical processes and the other material dealt with in the lessons. There are two main reasons for this:

1. At this stage of the course abstract thinking still proves difficult.
2. The students are confronted with technical information and material which is completely new to them.

For these reasons it is necessary, when starting a new subject, to make sure that the student understands exactly what it is that is being talked about. In such cases photos, slides and charts allow him to actually see what he is learning about. Such methods greatly facilitate the students' thought processes - looking, watching, comparing. This can then be followed up, without any intermediate stage, by a higher stage in the process of abstraction namely, symbolical presentation or the presentation of principle sketches. After explanation, this step paves the way for the next higher level of abstraction which is the mathematical generalisation (whereever possible) in the form of equations.

This means of inductive presentation seems the most suitable to us. Expressed in philosophical terms, training progresses from the Individual via the Particular to the GENERAL!

The process is continued after the lecture, discussion or seminar by presenting the students with the object in question (or with extremely true-to-life models of the object of the lesson) and by carrying out exercises.

Thus from the Abstract, we progress back to the concrete machines or, from the General to the Individual, the Particular. This method allows us to realise various didactic principles:
- Concrete and Abstract
- Clearness and intelligibility
- Consolidation and permanent acquisition.

4. Specialization

The 3 1/2 year long general training, which all students go through is followed by a course of specialization. This course which lasts one and a half years provide the student with a more comprehensive training, allows him to specialize in a subject of his own choice and requires him to prepare a project in detail on a particular subject which will count towards his final examinations (M.Sc.). The latter is the first scientific task which the student is required to produce independently and demands that each student has his own specific work schedule. As such, this part of the course is a reflection of the success of the course itself.

For students who specialize in farm machinery, the work schedules include both sections of the theoretical course and practical work in socialist enterprises. We attach considerable importance to the practical work as this gives the student the opportunity of dealing with tractors and agricultural machinery over a relatively long period and of getting to know the function, capacity, efficiency and potential of the tools of his trade at first hand. In this way, the students powers of judgement are improved and he becomes well-versed in dealing with the problems involved in agricultural technology. It is particularly important that discussion takes place at every stage of the course. This discussion should involve both the students and representatives from the various enterprises so that each process and question is adequately explained and so that the student is amply aware of the factors and difficulties in question. The evaluation of practical work should also take place in such communal groups where problems and questions still outstanding can be discussed once more.

The periods of practical work also form the basis for collecting the material which is to be used in the long written projects. We usually allow between three and three and a half months for preparing these projects which must then be discussed and defended by the students in front of a panel made up of scientists. Depending on the students themselves and on the research subjects agreed upon, the following subjects for specialization in the field of farm machinery usually prove the most popular:

- Maschines for harvesting threshing crops.
- Use of tractors

After successful completion of the course, it is possible that some of the students may be allowed to prepare a doctoral dissertation on farm machinery.

5. Summary and Conclusion

Our methods of teaching and of putting the material over to the students have proved successful - in this respect. We see our main task as being to teach the subject in such a way that the student learns to understand and grasp concepts and interrelationships and that his ability to think is developed during the course.

From experience, we can say that the following factors must be given pride of place in education and training:

- Teaching methods and material must correspond to the latest and highest scientific standards
- The quality of lessons can and must always be improved
- Certain subject must be chosen and concentrated on and this choice must depend on the structure and make-up of the study groups
- Written teaching material must be carefully developed
- Full use must be made of audio-visual aids
- Members of staff must be fully qualified.

Bibliography

/1/ Course curriculum for the training of foreign students in tropical and subtropical agriculture 1974.

Mechanization of Crop Production in State Farms of the Reclaimed Land in the Arab Republic of Egypt with Special Reference to the Agrarian Sector Mariut

by G. Menzel

1. Introduction

Mechanization of crop production is an important factor to exploit great resources in increasing production in agriculture of developing countries too.

One of the eminent examples in this connection are the mechanized state farms in the newly reclaimed land of A.R.E. Between 1969 and 1972, the author has had the opportunity of working in the sector Mariut which is one of the youngest and most mechanized of the above mentioned state farms. For this reason, the objectives of this lecture are aiming to declare the main experiences of mechanized crop production in Mariut.

2. Mechanized crop production in Mariut

Immediately after the revolution in 1952, the Egyptian Government decided to extend the arable land by reclaiming the desert and other waste area. In order to guarantee a complete and successful implementation of the land reclamation programme, a system of state organizations was created. Besides other state institutions, the General Organization for Utilization and Development of Reclaimed Land (EAUDRL)[x] was founded in 1966, being responsible for the development of more than 300,000 ha (1972) newly reclaimed land. EAUDRL is the headquarter of 11 agricultural sectors (state farms), which are situated within the Nile Delta and along the Nile Valley. Such a state farm has an average size of about 20,000 ha.

---

[x]Egyptian Authority for Utilization and Development of Reclaimed Land

The Mariut Sector lies on the edge of the North-West Delta with an area of approximately 18,000 ha. It is subdivided into 9 farms, each with an average size of 2000 ha. The soil is rich on calcium carbonate (lime) from 20 - 52 % and is recognized as one of the most fertile soil types of the A.R.E. Irrigation work depends on the water discharge of the Nile-Nubarea-Canal. In order to lift the water up to 30 m to the end of the sector, 4 main pumping stations were constructed. The reclaimed land of Mariut is subdivided into about 9000 plots, everyone with a size of about 2 - 2.5 ha. Cultivation started on 2000 ha land in 1966 and were extended step by step up to 16,000 ha with the following crop rotation: 14 % grape, 51 % fodder (alfalfa, egyptian clover, etc.), 18 % cereals (wheat, barley, etc.), 1 % oilseeds, 6 % vegetables, 4 % soil improvement and 6 % rented area.

18,500 people are living here in 18 villages, about 700 farmers of them with their families (newsettlers) and 3000 employees of the sector Mariut.

The agricultural sector Mariut (like the other one's) is a state enterprise serving as a pilot scheme of modern, mechanized production in crop growing and animal husbandry in order to encourage a progressive development of agriculture in Egypt. Introduction and development of mechanization in agricultural production was conceived as one of the most important steps to increase production and labour productivity. For that reason the Egyptian government has supported many forms of scientific-technical cooperation between the EAUDRL and the competent State Organizations in socialist countries. In spite of many problems, mechanization of agriculturel production has been proved as the main pillar of a successful development of the state farms, especially of the Mariut sector. In the annual report of 1972, the EAUDRL outlined the following fitting remarks on this subject: "The organization is applying the latest scientific and technological results and up-to-date technologies in order to employ tractors and agricultural machines in an increasing manner.

For that reason it was possible to reduce seed expenditures to 50 % and to increase net production by 20 %. The mechanized state owned agricultural sectors Mariut and North-Tahreer are good examples of how mechanization has been succesfully implemented in agriculture. It has enabled us to reduce expenditures by 6.5 Egyptian Pound per Feddan (1 Feddan = 0.425 ha) and year and to solve the problem of the lack of seasonal workers".[x] (8)

The following table indicates the high level of mechanization of agricultural production in Mariut sector compared with the same average level of all other sectors (EAUDRL):

Table 1 Mechanization in agriculture (%)

Operation	Mariut Sector	EAUDRL
Ploughing	100	100
Drilling	85	65
Plant protection (Spraying)	100	80
Threshing	100	100
from it combine harvesting	100	55

Source: Statistical Department of EAUDRL and of the Sector Mariut, October 1972. (unpublished)

In 1967, when the technical process of reclamation of the former desert was completed and the initial cultivation of partial areas was begun, the central authorities of EAUDRL started to develop the sector Mariut as a model farm of mechanization. Thus, from the very beginning it was decided to form the structure of crop production by concentrating on the mechanized cultivation of non-traditional crops like grape, oilseeds, etc. This is the only way in realizing an economical return according to the high amount (about 360 Egyptian Pound per Feddan) of investments by the state. (1)

With this target in mind, more than 2000 ha of grape plantations have been planted by 1972. Grape production shall be a basic element that's why the area is to be extended to about 7000 ha in the forthcoming period. It means that one third of the Mariut area will be planted with grapes. The technology of grape production is based on modern scientific and technical results, so that many work operations like weeding, spraying, etc. are almost mechanized. Thus, for example, the distance between the rows has

[x]In the state farms there is a great lack of seasonal workers from May - June and from September - October each year, which are periods of peak work in agriculture.

been set at 2.40 m in order to facilitate operating with special tractors like the Bulgar caterpillar, the multipurpose carrier GT 124 (G.D.R.) etc.

Introduction of special crops[x] such as sunflower, flax, beans, peas, watermelons, seed production of alfalfa, etc. is decided too and shall cover an area of 7000 ha. The final third of the Mariut area shall be cultivated with fodder crops like alfalfa, egyptian clover, being the basis for the development of milk and beef production to more than 10,000 heads of cattle in future. Gradually an increase of the above mentioned main crops will be combined with a process of specialization and concentration within the sector and at the level of the 9 farms. By this way it becomes increasingly possible to apply tractors and agricultural machines for work, which were usually carried out by hand like fertilizing, weeding, canal cleaning with chemical and mechanical (excavator) means, haymaking, etc. This extension of mechanized work is going to be typically, although it differs from crop to crop as can be observed in Table 2:

Table 2 Mechanization costs for selected crops in the Sector Mariut 1971/1972 (figures in LE per Feddan)

Crop Cost item	Grapes	Sun- flower	Maize	Wheat	Alfalfa- Seedprod.
Mechanized work	7.00	10.00	10.00	6.50	5.00
Manual work	25.00	4.00	4.00	2.00	0.85
other basic costs (seed, fertilizers etc.)	28.00	8.00	12.00	7.00	4.50
total basic costs	60.00	22.00	26.00	15.50	10.35
from it cost of mechanization in %	11.70	45.40	38.50	41.90	48.35

Source: Statistical Department, Mariut, October 1972. (unpublished)

It is necessary to achieve the highest feasible level of mechanization for each crop so far as the suppositions can be created like the qualification

[x]for export and other urgent demands of the economical development of A.R.E.

of the workers, tractor operators, engineers, the level of planning and management, the technical equipment available etc.

This target also includes an increasing extend of mechanized operations in processing, storing, grading and cleaning of agricultural products for export as well as for domestic consumption. As far as quantity is concerned, the equipment available for production is nearly sufficient in mechanizing crop production.

This is confirmed in Table 3, containing figures on the number of farm machinery in Mariut.

Table 3 Number of tractors and farm machinery in the Mariut Sector

Farm machinery		1970	1972	Perspective requirement
1. Tractors				
Total number		371	341	approx. 300
Number available for operating		146	122	approx. 240
% available for operating		39	36	approx. 80
2. Combine harvester				
Total number		28 E177	18 E177 8 E512	approx. 10 E177 approx. 12E512
3. Selected agricultural machines:	Ploughs	25	40	approx. 40-50
	Leveller	10	15	approx. 30-40
	Seed planter	12	30	approx. 25-30
	Sprayer	50	60	approx. 40-50
	Baler	15	25	approx. 20-25

Source: Statistical Department, Sector Mariut, October 1972 (unpublished)

According to the figures of table 3, it can be concluded, that the number of tractors and selected agricultural machines are sufficient in order to meet the present and perspective requirements. On the other hand, the regular technical capacity can't be utilized for the very low percentage of tractors available for operating, adversary effecting the increase of production and labour productivity as well as profitability. The main reasons for that situation are the following:

- organizational and technical shortcomings as regards operating and utilization
- poor quality of repair, especially a low standard of service and maintenance;
- lack of workers and qualified employees such as mechanics, tractor operators and engineers;
- difficulties in the field of the supply of spare parts, tools, etc.

3. Training, service and effective operating - important suppositions for agricultural mechanization

Training of the technical staff, establishment of a comprehensive system of service, maintenance and repair and well-organized operating of tractors and agricultural machines shall be explained in the following, because they form the nucleus of experiences, gathered in succesful mechanization of crop production in Mariut.

3.1. Training

Economical use of tractors and agricultural machines depends mainly on training technical staff. Only a few technical engineers were theoretically educated in agricultural technical subjects whilst many of the workers who should work as operators, mechanics, welders etc., were illitirates. Employing such workers without basic training in a practical way will increase the shortcomings and problems as mentioned above.

For that reason practical training courses were organized which took place on the fields, in workshops etc. being a short-time training in practical work with tractors, combine harvesters, seed planters, welding machines etc.

But soon, it was necessary to establish a net-work of training centres in conformity with the increase and modernization of agricultural techniques in order to introduce the education of skilled worker, long-term training of operators, mechanics etc. and to improve the conditions for carrying out short-term courses for adult training.

Table 4 Training centres of the General Organization of Utilization and Development of Reclaimed Land

Name of the sector with training centre	Kind of long-term education of skilled workers	Duration (years)
Gianaclis	tractor operators	0.5-1.0
North-Tahreer	skilled workers	2.0
Mariut	agricultural mechanics	2.5
East-Delta	agricultural mechanics	2.5
Middle Delta	agricultural mechanics	2.5

The following results, gained in the training centre Mariut, are intended to get an idea of the activities of those training centres:
Between 1968 and 1972 about 80 young Egyptians completed a two and a half years apprenticeship. They had obtained certificates as skilled mechanics and were delegated to repair tractors and agricultural machines in spezialized workshops, service stations and combine harvester groups of the sector Mariut, North-Tahreer and Nahda.

Apart from the education of apprentices, many short-term courses took place every year. In this way 60 farm directors, operating engineers and other members of the supervisory staff, 28 operators of combine harvester and 120 tractor operators were trained.

3.2. Service, maintenance and repair

A well organized and effective system of service, maintenance and repair for all types of tractors and agricultural machines represents an important supposition in order to make use of the advantages of mechanization. There is such an unified system established in each agrarian sector of EAUDRL consisting of:

- service stations in most of the farms
- workshops for medium repair work
- spezialized centre workshops (general repair of engines, fuel pumps, crankshafts, hydraulic pumps etc.)

- main stores of the EAUDRL in Alexandria, South-Tahreer and Cairo (mainly for spare parts, assemble groups, tractors, and machines, which are imported from abroad)

The service stations are technical centres of an agricultural farm, which has a cultivation area between 1500 ha and 3000 ha. The tasks involve operating, garaging, service, maintenance and small-scale repair of the number of tractors, lorries, cars and agricultural machines, which were permanently alloted to the farm staff. But the outlined tasks could only be partially fulfilled because of the shortage of operators, mechanics and other problems.

In order to overcome those problems, which were hindering the development by giving rise to unsatisfactory performances, a special inspection-maintenance-service was set up. Two workshop cars were equipped with an excellent staff (engineers, mechanics, tools, welding machines etc.) in order to check, assist and train the workers (operators, mechanics, foremen) of the service stations. This new method has proved succesfully and was recommended for reapplication in all sectors of the EAUDRL. The workshops are another important chain of the above mentioned system. After completion of the initial development stage with its problems and difficulties (building, assembly of machines, choice and training of the staff, current and water installation, purchase of spareparts etc.) it was possible to specialize the activities of each workshop. The former repair work, carried out to all types of tractors and machines in each workshop, was characterized by a very low performance and labour productivity. Therefore specialization in repairing was tackled according to a programme, which was elaborated by the EAUDRL and valid for all the 11 sectors. The following chart of the technical department in Mariut contains the number of workshops (4) and their different tasks as well as relation with the specialized workshop (engine repairing) of North-Tahreer.

Structure in the Field of Mechanization

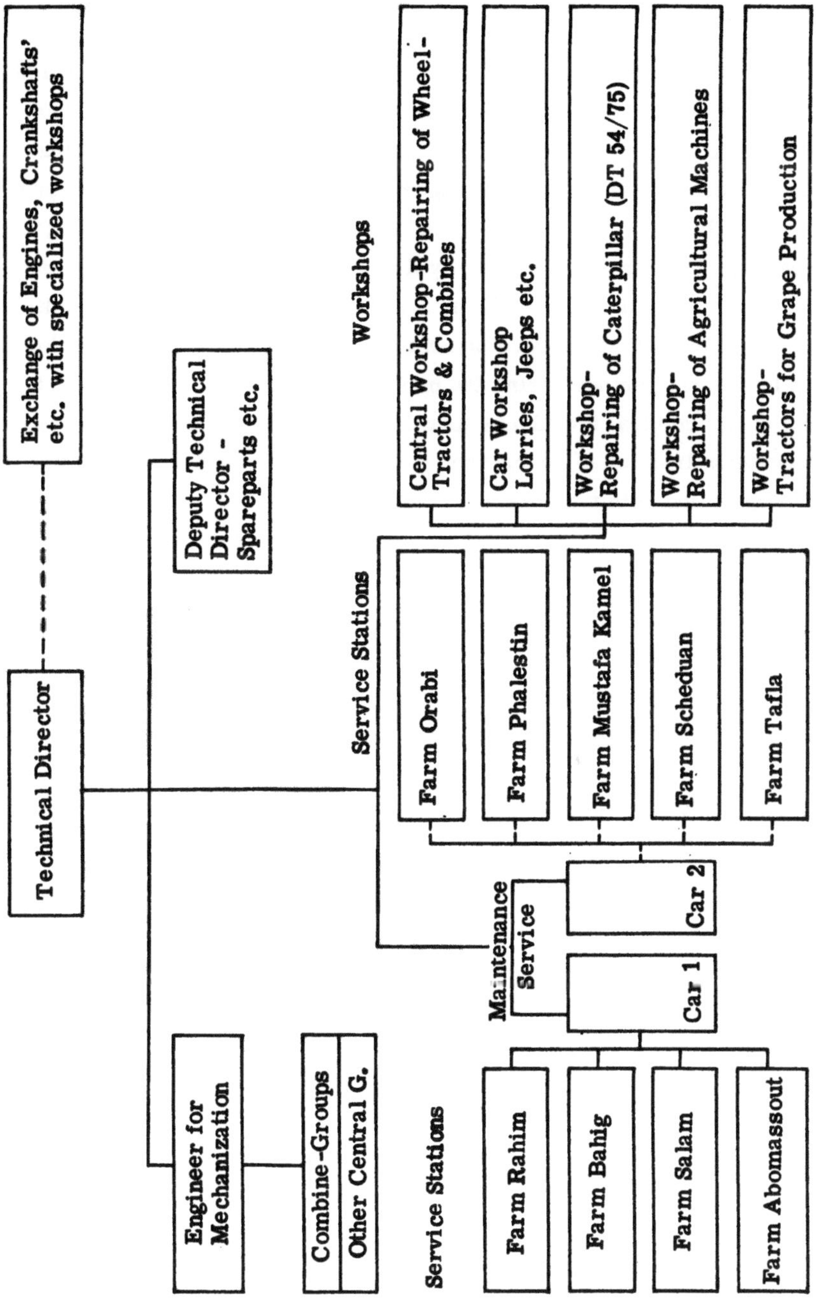

3.3. Effective operating

Succesfully mechanized agricultural production depends rather on a high standard of planning and management to be achieved by the headquarter of EAUDRL, sectors, farms and workshops. We have to recognize that the importance of planning and management steadily increases as the process of specialization, concentration and cooperation of social work is going to proceed. This process is directly linked with an increase in the use of tractors, agricultural machines, fertilizers, weedkillers, pesticides etc.

This means that the development of crop production has to be planned on long-term periods, being the only way for an economical use and planning of agricultural techniques (purchase, completion of machine systems, long-term training of specialized workers and supervisory staff etc.).
But operating with tractors and machines in the most efficient way depends also on short-term planning. According to the experiences gained in Mariut, it is very essential to prepare seasonal working programmes (cultivation, harvesting etc.) as early as possible by the management-team and the workers involved. The increasing participation of operators, mechanics, etc. is very important, if the respective programmes and plans of mechanization (perspective, year and saison) shall be really implemented. This is the only way to create the subjective conditions being of great influence on wether the confirmed targets of production and income can be achieved or not. Another experience has tought us that management and supervision should be given in accordance with the requirements of mechanization. The latest stage is shown in the chart of the technical department of Mariut.

As already mentioned, it is possible and necessary to operate with the modern technique in applying the best international methods of planning and management in state farms and cooperative farms. This applies especially to the state farms and cooperative farms of the socialist countries. One of such methods, is complex or group work of combine harvesters, balers, caterpillars and other types of tractors. Since 1967, combine harvesters and balers were operating in several complexes being directed by the general manager of the Mariut sector, respectively the operating engineer. Wheat, barley, alfalfa (seed production), egyptian clover, rape, beans and peas have been harvested by combines or were stationary threshed depending on the harvesting method used. In 1972 eight combine harvesters type E 512 (high performance etc.) were available for harvesting the above-mentioned crops including sunflower and maize (2).

Table 5 Harvesting of crops with two E 512 combine harvester complexes (each with 4 combines) - Mariut sector, 1972

crop	total area harvested in ha	harvesting month
rape	149	May
flax	255	May·
barley	385	June
wheat	1, 533	May/June
kale	17	July
Egyptian clover	145	June/July
alfalfa	839	August
sunflower	94	September
maize	1, 105	October/beginning of November
Total	4, 522	nearly 7 month operating time

Source: Statistical Department and Technic Department, Sector Mariut, November 1972 (unpublished)

As table 5 indicates, a total area of 4, 522 ha had been harvested by 8 combine harvesters of the type E 512. Each combine had achieved a yearly performance of 565 ha, doubtless an excellent and very economical result. The continuous improvement of the planning, management, and qualification level in the field of mechanization has decisively contributed in increasing labour productivity and economical utilization of modern farm machinery as the figures of the following table are demostrating.

Table 6 Economical Results of operating with combine harvesters in Mariut (1969 - 1972)

Year	harvested area ha	type of combine harvester	number of combine harvester	Performance combine daily ha	yearly ha	waste time (in comp. to the total work time) %
1969	2,213	E 177	26	3.6	85	48
1970	1,630	E 177	24	4.5	68	58
1971	1,966	E 177	20	3.6	98	41
1971	1,682	E 512	4	8.7	423	20
1972	1,488	E 177[x]	6	2.6	248	35
1972	4,522	E 512	8	7.7	565	23

[x]E 177 - only stationary operating for threshing of beans and peas

Source: Statistical Department and Technic Department, Sector Mariut, November 1972 (unpublished)

The waste time - expressed in percentage of the total harvesting time during a season - was systematically reduced both by an improvement in management and operating with modern combine harvesters. At the same time some other important results were achieved by operating with modern combine harvesters such as:

- considerable decrease of harvesting losses to about 3 - 5 %,
- ease of operating and adjusting by the operators,
- reducation of harvesting time has facilitated in improving the quality of recultivation and in better keeping of the best datas of cultivating all crops.

In order to generalize the advantages of complex work with high performanced tractors, agricultural machines etc. the following mainpoints shall be concluded:

(a) Each complex respective group should be equipped with appropriate tractors, machines, tools etc. which can be really applied and economically used by the head of the complex. Thus, for example a complex of combine harvester E 512 should consist of 3 - 5 combines, 1 workshop car, 2 lorries (transport of grain etc.), 1 car for the head and 3 - 5 high pressure pick-up balers.

(b) Choice and training of the whole staff as early as possible. One should rely as far as possible to nominate engineers, operators etc., who had worked successfully in the same job the previous year.

(c) Payment of wages per hour respective time is very unsuitable, because there is no stimuli in utilizing the real performances of agricultural techniques with high value respective capacity. Extra payment (bonus etc.) should be granted in case, the daily standard of work (quantity and quality) is exceeded.

(d) It is very important to elaborate a working programme for such a complex for not less than 2 months, before the harvesting season of summer or winter crops - will begin. But many other preparations have to be made earlier in order to utilize the great advantages. For example, as already mentioned the plots of Mariut have a size of 2 - 2.5 ha on average. In order to make full use of the capacity of a complex of combine harvesters E 512 (4 pieces) connected areas of not less than 50 ha a day are needed. Therefore the competent manager has to decide to cultivate crops, destined to be harvested by combine harvesters, not less than in connected areas of 100 ha (equal to 40 - 50 plots in Mariut sector). Furthermore it should be also mentioned that the irrigation system must be modificated in accordance with the requirements of mechanized agricultural work (2, 6).

Summary

Mechanization of agricultural production is very essential and advantageous in state farms and in developing newly reclaimed land. The results and experiences, gained in the development of the sector Mariut and other state farms of the EAUDRL in A.R.E. must be utilized.

A good progress in mechanization of agricultural production depends mainly on:

- comprehensive and practical short and long-term training for the whole staff, particularly in the field of operating with tractors and agricultural machines according to modern technologies;

- elaboration and consequent implementation of an effective system of service, maintenance, repair and purchase of spare-parts or other materials. It should be based on a limited number of types of tractors and agricultural and other (equipment for workshops etc.) machines in order to prevent difficulties;
- well-organized management and planning of operating with agricultural techniques and the work of the staff involved.

Bibliography

(1) Badran, O., Report of the Ministry for Land Reclamation of the Arab Republic of Egypt; Egyptian Gazette, 23.7.72, p. 2.

(2) Glanze, P., Report on the use of the combine harvester E 512 by the Testing station of VVB Land- und Nahrungsgütertechnik in the Arab Republic of Egypt. Leipzig, March 1974 (unpublished).

(3) Menzel, G., The development of the Mariut Sector 1969-72, Problems and Conclusions; Dec. 1972 (unpublished).

(4) Michalski, K.-J., Agricultural cooperative societies in developing countries of Africa and Asia. Akademie-Verlag, Berlin 1973, p. 271/272.

(5) Michalski, K.-J., Pfeiffer, A., Economical problems in the noncapitalist development of agriculture in the developing countries. Protocol of the Scientific Conference of the Institute of Tropical Agriculture and Veterinary Science. 18/19.6.1970, p. 39/40.

(6) Peipp, L., Mechanization problems in agriculture in the developing countries. Protocol of the Scientific Conference of the Institute of Tropical Agriculture and Veterinary Science, Leipzig, 18/19.6.1970, p. 107/110.

(7) Ram Das, Feasibility Study on the Establishment of Agro-Industries in the newly reclaimed sectors of North-West-Delta, Mariut and North-Tahreer, Arab Republic of Egypt, Ministry of Land Reclamation. June 1972 and Evaluation and Action Research UN/OTC/A.R.E. 71.

(8) Anonymous, Egypt - The New Land. Brochure on the development of the Egyptian Organization on Land Development. Cairo, March 1972.

The Economic, Social and Political Impact of Agricultural Modernization on the Latifundio Agricultures of Latin America

by Ernest Feder

During the last few years, great efforts have been made to "modernize" Latin American agriculture and to demonstrate that there is not in fact incompatibility between an extreme concentration of ownership and intensive agriculture. These efforts are now most likely to be redoubled because of the world-wide shortages of staple foods. On these two issues I shall devote the remainder of this paper.

Efforts by individual landowners and by national governments through appropriate government programmes to increase the output of agricultural commodities by more efficient methods are by no means new, although in Latin America these efforts have been extremely modest until the 1960's.

As far as the establishment of modern farm enterprises using up-to-date methods of cultivation (including modern technology) is concerned, this took place in the past usually in one of the following ways:

(a) through the plantations, producing for export, usually with heavy foreign capital investments. Once established however, these enterprises have continued in many cases to be operated along the methods used at the time of their establishment, meaning that once the capital investment was made, the aim of the operation was to recuperate the capital investment in the shortest possible time and to "mine" the land for maximum profits. This explains why many - though of course not all - plantations are now relatively unproductive and cannot compete effectively in the world markets.

---

[x] Epitome

(b) though the initiative of large and smaller progressive landowners interested in managing their property at a high level of efficiency. Usually these efforts are shortlived and end with the disappearance of the owners.

(c) through rich urban investors becoming estate owners for the purpose of evading taxes or acquiring social status and prestige or as a hobby, or in order to benefit from unusually favorable market conditions and government subsidies (through favorable credits, price guarantees and the like).

At times, such efforts have resulted in local transformations of the output pattern and the socio-economic structure - as was the case for example with coffee in Brazil (the "coffee kings") or sugar or cotton in a variety of Latin American countries.

(d) through the settlement of cultivators from developed agricultures, such as Germany or Switzerland emigrating to the Latin American continent, or more recently from Japan. These settlements are often encouraged by national governments on the pretext that the new settlers would have a "demonstration effect" on the local farm people. In many, but no means all, instances these settlers were highly successful, but in practically no case has there ever been a "demonstration effect", for the simple reason that the conditions under which these settlers obtain the land, credit, machinery or fertilizers were totally different from the conditions under which most local producers, particularly the small ones, could operate their farm and the national governments would never consider granting the bulk of the local producers similar facilities. (1)

None of these ways has however affected on any large scale the performance of Latin American agriculture dominated by the unproductive latifundios.

There have also been sporadic and periodic government programmes to stimulate output. In other parts of the world, these programmes come under the heading "food drives" and the like, i.e. the intensification of the cultivation of food crops. But in Latin America, the programmes were less systematic and were focused more often than not on specific commodities which need not necessarily be food products for domestic consumption.

A favorite - if somewhat amusing - type of programme has been for the increased production of livestock - Latin America's mosts inefficient agricultural industry - for the purpose of increasing exports. It is amusing because the governments were usually unwilling or unable - on account of the livestock growers' indifference or hostility - to accompany these programmes by schemes to improve the health and sanitation conditions necessary to meet the requirements of importers. To implement such schemes would have meant strict government controls which the producers are totally unwilling to accept. More indirectly the governments have supported production programmes by channeling extra resources (credit, technical assistance, research etc.) to specific commodity associations (wheat producers', etc.) which are always associations of large estate owners. But under the conditions under which such support is given, it could hardly result in a significant modernization of the farm enterprises because the support is not tied to an obligation of the producers to modernize their farming operations. (2)

Suddenly in 1959/60, with the advent of the Cuban revolution and land reform, modernization strategies have become much more important. (3)

The points I would like to stress in the following paragraphs are the following:

(a) modernization of agriculture has become more openly a political tool to defend the existing political and socio-economic structure and with it the dependency of many countries in Latin America on the industrial nations, particularly the United States;

(b) modernization has become more directly a concern of national governments and industrial nations, particularly the United States, rather than a matter of private initiative;

(c) what modernization implies;

(d) whether modernization can overcome the traditional obstacles to raising output significantly;

(e) what the consequences of modernization are likely to be.

In order to understand the origin and aim of the new modernization drive in Latin America, it is necessary to summarize briefly the impact of the

Cuban revolution on Latin America's ruling classes and the rural proletariat and on the government of the United States. It is well known that the immediate reaction to Cuba's socialist revolution and land reform was the Alliance for Progress (1961) which in turn was a response in part to the growing peasant unrest in practically all Latin American nations as a consequence of Cuba's land reform. The aim of the Charter of Punta del Este, which was the international agreement embodying the functions and aims of the Alliance, was to avoid, practically at all costs, a repetition of the Cuban events by paving the way to legal, orderly and peaceful agrarian reforms which were to be carried out by the existing governments. This was equivalent to saying that these reforms were to be carried out by landed elite itself (the chief intended "victims" of the reforms) since the political power of the established governments was derived exclusively or, in the worst of cases, to a very large extent, from the landed property owners. It is obvious therefore that the land reforms carried out by these governments had one principle aim: namely to pacify the peasants who in the beginning of the 1960's had founded powerful organisations in such countries as Brazil, Colombia or Venezuela and initiated large-scale peasant movements demanding more land and the abolition of the latifundio system. (4)

Through the Charter, the United States offered to tie financial assistance to the implementation of agrarian (and other) reforms, and as a consequence practically all Latin American governments passed new land reform legislation and set up (at times impressive) land reform institutes. However, the principle of "tied financial assistance" was never effectively carried out, and as subsequent events showed no real land reforms were carried out anywhere under the Alliance. (5) To a large extent, if not perhaps exclusively, this was due to the fact that simultaneously with the Alliance, which appeared at first to inaugurate an entirely new, progressive era in the hemisphere, the military opened up an undeclared war on the peasant organisations which was so successful that by the mid-1960's all peasant organisations had been destroyed through violent and non-violent means. (6) This set also en effective end to political pressures in favour of reforms.

Peasant pressure for more land could be silenced, at least for the time being, by killing or jailing actual or potential peasant leaders and keeping the peasants in disarray. But this strategy did not solve the persistent problem of inadequate food supplies and the inability of the estate sector, which controls most of the farm land and practically all of the good quality soils, to improve its performance. While it appeared relatively easy to control peasant discontent, the need to import staple foods threatened foreign exchange reserves and could also become a disrupting factor for social peace in the rapidly growing slums surrounding all major cities. Social unrest in the cities could become as important a threat to the "establishment" as the peasants' claim for land. Hence the strategy of the Latin American governments and the United States was twofold: to modernize the latifundios and to set up at least a nucleus of highly productive, capital-intensive enterprises in the estate sector producing staple foods (e.g. rice, wheat) and products destined for exports (e.g. cotton, sugar) both to supply domestic markets and as foreign exchange earners; and to push agriculture more rapidly into the virgin areas. It is my judgement that those programmes were undertaken very consciously and systematically. The land reform institutes continued their platonic land reform projects to pacify the peasants (at a low cost for the governments) and at the same time the latifundio sector was enormously strengthened by a substantial injection of capital, technical assistance and other advantages - a demonstration of the evident economic truth that costs are no object when it comes to maintaining the status quo. One of the political aims of this strategy was to demonstrate also that large estates can be efficient and productive and hence an indispensable element of Latin American agriculture and thereby counter the argument that land reforms are necessary because the big estates do not utilize their land or utilize it inefficiently.

The strengthening of the latifundios took on various forms. The most important were the massive importation (or, in some cases, the local assembly) of mechanized modern equipment (adapted exclusively to large-scale farming and totally inaccessible to small operators) to be sold to rich estate owners under very advantageous conditions; massive credit facilities; the distribution of improved seeds and fertilizers and improved marketing, including storage facilities and price guarantees. This process was aided significantly by inflation as it encouraged urban investors to buy up land in old established farming communities or in newly opened virgin areas, and in fact these urban capitalists with "modern ideas" have led the way all along in this modernization scheme throughout the continent. No less important were the substantial funds flowing into the Latin American continent for new infra-structures or for improving old ones (e.g. irrigation schemes and roads), the benefits of which accrued almost exclusively to the big landlords.

The United States and other industrial countries have ever since exerted strong pressure directly or indirectly (e.g. through the farm equipment industries) to bring the "advantages" of modern farming to Latin America, for example by granting financial aid with the specification that the funds be used for the importation of mechanized and other farm equipment. It is to be noted however that with the exception of Mexico, where modernization has started at an earlier date with United States initiative and assistance, the "Green Revolution" (7) has not been pushed in Latin America as it has been in Asia for example, for the plausible reason that once the peasant movements had been destroyed, the political risk of land reforms or revolutions had abated, and the simpler process of transfers of equipment and other inputs for use only by large-scale producers proved to be adequate to strengthen the latifundio-sector.

In summary, the modernization of the latifundio sector means the introduction of better technology (including the mechanization of farming operations, better farm inputs, better management) without changing the basic socio-economic and political structure in agriculture. It implies that technology is being used as a tool not only to improve the output performance, but also to "solve" the social, political and economic problems in agriculture. It is in short the technocrats' way of attacking the agricultural and agrarian difficulties in rural Latin America. Hence under the conditions under which it is being implemented, modernization is essentially a counterreform programme even if - or better: particularly if - it is successful in rising the productive capabilities of the now partially modernized latifundio sector.

As a result, it would be logical to expect that the short-run benefits in terms of a likely increase in food and fibre output would be offset and perhaps much more than offset, by the likely difficulties arising out of the negative impact of the programme.

To these I shall now turn my attention.

It is very likely that the systematic creation of a nucleus of modern capital-intensive latifundios will enable Latin nations to become "self-sufficient" with respect to staple foods consumed domestically and increase the production of exportable products. However as I see it, this expansion of output will encounter various limitations and it should be noted in passing that according to available estimates, the food situation in Latin America has not improved significantly, if at all, over the last 10-15 years, though it would appear that without the modernization efforts this situation might well be worse than it is today.

In the first place, we have to ask ourselves: Self-sufficiency at what level? It is evident that in the capitalist economies of Latin America, the commercial output of food products adjusts itself, within certain limits, to the purchasing power of the domestic population with respect to staple and other food products consumed at home, i.e. to the "effective demand" and to the ability of the world markets to absorb increased quantities of food and fibre. Since the purchasing power of a large section of the consumers is very low (and in fact declining because of their declining real incomes), the increase in commercial production of domestically consumed foods must be in step with the increase of that section of the population which has the ability to buy - that is, grosso modo, with the increase in the size of the middle classes and the small privileged section of the industrial workers. If output were to increase beyond this limit, prices would fall and production would be curtailed, unless the "surplus" could be exported. In no case, however, would the "surplus" benefit those sections of the population with inadequate purchasing power, since there are no mechanisms in these economies to distribute food in accordance with needs: This would be inconsistent with the basic principles of a capitalist economy. If prices of the products fall, producers would have to carry the losses or the government would have to step in wherever the products are price-supported. It is clear that the existence of large-scale capital-intensive commercial latifundios introduces an element of instability and uncertainty in the food production and marketing picture. The important thing is however that purchasing power sets a real limit to the growth rate of commercial food production. This is no doubt one of the causes of the slow increase in commercial agricultural output which we have witnessed in the past. Hence without a radical change in the distribution of wealth and income it can hardly be expected that in the foreseable future the performance of the agricultures of Latin America can be radically improved, except if export markets could somehow absorb the food "surpluses".

The recent oil crisis which resulted in considerably higher prices for gasoline and other petroleum products will in all likelihood intensify the limitations on the demand side for food and fibre as the increase in the costs of production of capital-intensive, large-scale commercial farm enterprises will be passed on to consumers in the form of higher prices and thereby narrow the domestic and even foreign markets. On the other side the limited "self-sufficiency" programme tailored to "effective demand" which I have described above, will not help to improve the foreign exchange situation because even if imports of staple foods are totally eliminated, the increased costs of oil imports would prove to be a drain on foreign exchange reserves for most Latin American countries. Therefore Latin America is faced with a real dilemma, given the institutional context within which it must operate: The modernization programmes which are designed to secure more food on one side and on the other to secure the survival of the existing anti-social land tenure system, are bound to run into increasing difficulties.

These arguments may seem to be rather speculative, were it not for the spectacle presented by the development of Mexican agriculture over the past 20-25 years which demonstrates the inherent contradictions of capitalist-style agricultural development. Mexico is the cradle of the "Green

Revolution". (8) The development and large-scale use of high-yielding varities of wheat, corn (maize) and sorghum made available exclusively to large-scale commercial growers particularly in the irrigation districts, became the cornerstone of a new latifundio-system in which capital intensive, highly mechanized farm enterprises predominated. In addition, there are important commercial producers of fruits and vegetables or cotton, exporting a substantial proportion of their output to the United States. Formerly a food importer, Mexico became an exporter of staple foods and other products as the rate of increase of agricultural output rose to unprecedented heights. Staple foods were exported although the diet of large and increasing sections of the Mexican population with low incomes became progressively more inadequate. By the end of the 1960's production increases began to level off and Mexico is now performing no better than other Latin American countries and its population is no better fed. (9)

Another serious limitation to modernization is the traditional attitude towards land utilization and farm management. This implies that the modernizing efforts of a few individual producers and governments are bound to be offset by the continued wasteful land use and farm management practices of the majority of the landowners. The intensification of the use of land is limited to a small and only slowly growing number of producers - not a planned programme to lift the entire agricultural plant up to higher levels of performance.

The modernization programmes are not development plans in the proper sense of the term, but a piecemeal approach to secure a better supply of specified food products on specified privileged enterprises. None of the Latin American countries has a systematic and well thought-out - in fact any type - of global land utilization and production plans for the better use of all physical and human resources and for creating massive employment opportunities in and outside of agriculture to raise the purchasing power of the masses, and if they had them, they could not enforce them within the prevailing institutional structure which glorifies individual initiative and private profits. For this reason the increased yields on the modern enterprises will be largely offset by decreasing yields on the traditional enterprises and the small-holdings and these trends will continue as long as there are still virgin areas available for future cultivation or until there is a revolutionary change in the land tenure structure and the entire socio-economic system.

References

1 The case of the Japanese settlers in Brazil is enlightening because they had the full support and almost unlimited assistance from the Brazilian and Japanese governments, which gave them a highly privileged status in Brazilian agriculture. None of this support has been forthcoming for the bulk of Brazil's local producers.

2 A typical case is the programme of the Brazilian federal government some 15-20 years ago to bail out the cocoa growners of Bahia, many of whom were supposedly facing bankruptcy. Enormous funds were allocated to producers willing to improve their plantation management and yields and renovate cocoa stands. The opposition of the powerful producers was however so strong that funds had finally to be allocated to growers on their own terms which meant that they used them to increase the capital value of their property (new buildings, stables, drying facilities etc.) without improving the declining productivity of their plantations.

3 A similar development occurred in other parts of the world, particularly those in which the "Green Revolution" was started.

4 The literature on these events is now very abundant. See for example R. Stavenhagen (Ed.) Agrarian Problems and Peasant Movements in Latin America, Anchor Books, New York, 1970.

5 In fact, in the few instances, in which governments attempted, or proclaimed their intention, to carry out land reforms, financial aid was actually withheld, as in the case of Brazil or Chile, and was forthcoming only after it became clear, as in Brazil after the military coup of 1964, that the governments did not take the land reform issue very seriously.

6 See E. Feder, Counterreform, in R. Stavenhagen op. cit. ...

7 The "Green Revolution" is a strategy of the United States (through the facilities of the Ford and Rockefeller Foundations) to distribute high yielding varieties of rice, wheat and corn to developing countries, which must be cultivated under optimum conditions and thus open the doors wide open to sophisticated farming equipment and other imports of inputs produced by the industrial nations. The major difference between the "Green Revolution" approach and the modernization programme implemented in Latin America is therefore the element of the distribution of high yielding varieties produced and made available to developing countries by the Ford and Rockefeller Foundations. The end effects of the two

programmes are however quite similar. For an interesting analysis of the "Green Revolution" see H.M. Cleaver, The Green Revolution and Imperialism, Monthly Review (New York) June 1972, pp. 86.

8 For an excellent article on Mexico's agricultural development strategies, see Cynthia H. de Alcantara, Die Geschichte der Grünen Revolution: Die Erfahrungen in Mexico, in Feder, Gewalt etc. op. cit., chapter 26 (in English: in Development and Change, The Hague, Vol. 5 no. 2, 1973).

9 For a brilliant study of Mexico's land tenure problems see Centro de Investigaciones Agrarias and CIDA, Estructura Agraria y Desarrollo Agricola en Mexico, 3 volumes, Mexico DF, 1970, soon to be published by El Fondo de Cultura Economica, Mexico DF.

Bei Fragen zur Produktsicherheit wenden Sie sich bitte an:
If you have any questions regarding product safety, please contact:

Walter de Gruyter GmbH
Genthiner Straße 13
10785 Berlin
productsafety@degruyterbrill.com